EVAPORATION OF WATER

With Emphasis on Applications and Measurements

Frank E. Jones

CRC Press
Taylor & Francis Group
Boca Raton London New York

CRC Press is an imprint of the
Taylor & Francis Group, an **informa** business

First published 1992 by CRC Press
Taylor & Francis Group
6000 Broken Sound Parkway NW, Suite 300
Boca Raton, FL 33487-2742

Reissued 2018 by CRC Press

Jones, Frank E.
Evaporation of water : with emphasis on applications and measurements / Frank E. Jones.
p. cm.
Includes bibliographical references and index.
ISBN 0-87371-363-X
1. Evaporation. 2. Evaporation—Measurement. 3. Evapotranspiration—Measurement . I. Title
QC304.J66 1991
530.4'27—dc20 91-18818

A Library of Congress record exists under LC control number: 91018819

Publisher's Note
The publisher has gone to great lengths to ensure the quality of this reprint but points out that some imperfections in the original copies may be apparent.

Disclaimer
The publisher has made every effort to trace copyright holders and welcomes correspondence from those they have been unable to contact.

ISBN 13: 978-1-315-89286-3 (hbk)
ISBN 13: 978-1-351-07196-3 (ebk)

Visit the Taylor & Francis Web site at http://www.taylorandfrancis.com and the
CRC Press Web site at http://www.crcpress.com

This book is dedicated to my lovely Christian wife, Virginia and to our talented children, Cynthia and Christopher.

He causeth the vapours to ascend from the ends of the earth.

Psalms 135:7

Preface

The loss of water from storage areas such as reservoirs, lakes, rivers, oceans, vegetation, earth, and irrigation conduits is a major concern of hydrologists and irrigation specialists. This loss, coupled with the relative scarcity of water in some areas, indicates the necessity for understanding the parameters and processes that comprise and contribute to evaporation.

This book reviews the literature pertinent to the fundamental processes involved in the evaporation of water. It covers many more practical areas as well. I hope it will be useful to a diverse readership including hydrologists, meteorologists, climatologists, irrigation specialists, civil engineers, chemical engineers, water conservation specialists, hydraulic engineers, agricultural engineers, agronomists, water resources specialists, geophysicists, fluid dynamicists, environmental engineers, and water storage specialists, as well as farmers and others who store water or use it for irrigation. The writing is intended to be simple and easily understood, and both SI units and English units are used in order that units not be a barrier for anyone. The intent of the book and many of the subjects covered are quite different from those of Brutsaert's excellent book.*

The book begins with an introduction in which the importance of evaporation is stressed. The volume of water in oceans, freshwater lakes, saline lakes, inland seas, rivers, soil, and the vadose zone is listed. The annual precipitation on the ocean and land areas of the earth is equal to the annual evaporation from the ocean and land areas, a very interesting and important conservation of water in the hydrologic cycle. The need for adequate knowledge of evaporation in the water balance and water management and the need for consistent measurement of evaporation and evapotranspiration are emphasized. The conservation of water in storage facilities, and attempts to suppress evaporation from them, is mentioned. Agricultural irrigation as the largest consumptive user of water in the United States, in which there are 60 million irrigated acres contrasted with 500 million irrigated acres in the rest of the world, is stressed. Loss of water by evaporation from water drops and

* Brutsaert, W. *Evaporation into the Atmosphere* (Boston, MA: D. Reidel Publishing Co., 1982).

attempts to stabilize aqueous fogs and atmospheric mists are mentioned, as is the application of the knowledge of water evaporation to other problems.

Following the introduction, there is a detailed treatment of the transport of water across the liquid water-gas (air and water vapor) interface including the following topics: kinetic theory of gases, evaporation and condensation of water, the principle of detailed balancing, absolute reaction rates, the concept of diffusional resistance, heat and mass transfer, flow in a fully developed turbulent boundary layer, evaporation by spurts, and evaporation of water in electric fields. Perhaps incongruously, the preparation of water is included in this chapter.

In Chapter 3, the controversial and occasionally misunderstood subject of evaporation and condensation coefficients is presented with many summaries of experimental investigations of these coefficients. The Hertz-Knudsen equation is given in simple forms, along with an example of the calculation of the evaporation coefficient using the equation. A discussion of the maximum rate of evaporation of water attempts to account for confusion in the use of the Hertz-Knudsen equation. The discussion of the evaporation of water drops includes the kinetics of the evaporation of droplets, evaporation from freely falling water drops, and the effects of insoluble films, charged water drops, vertical tunnel studies, and evaporation losses from sprinkler irrigation systems.

Surface tension, convection, and interfacial waves are combined in Chapter 5. The subjects covered are surface tension, convection, effects of waves, effects of monomolecular films, the effect of uniform rotation, effects of a progressive wave, liquid flow patterns, heat transfer and thermal structure in a boundary layer, flow patterns in an evaporating liquid layer, wind generated waves, and the effects of waves on evaporation. The measurement of surface temperature (an important quantity in the calculation of evaporation and condensation coefficients), temperature differences between water surface and the bulk water, temperature gradients, temperature of water in a film-treated reservoir, and water vapor distribution above a water surface are treated in Chapter 6.

Similarity, dimensionless groups, and wind tunnel experimentation are covered in a chapter on wind tunnel investigations of evaporation. The use of monomolecular films in attempts to suppress evaporation is treated in detail in Chapter 8 under the following topics: reduction of evaporation rate by monomolecular films, steps in the evaporation process, changes in the gaseous diffusion layer, changes in heat flux and near-surface temperature structure, alteration of surface temperature, heat and mass transfer, increase in temperature of water in a reservoir, resistances to evaporation, evaporation rates of film-coated water drops, stabilization of water drops, inhibition of evaporation from agar gel, effects of monolayers on the rate of evaporation of water and on the solution of oxygen in water, the effects of traces of permeable substances, and the reduction of evaporation from lakes and reservoirs.

Equations used to calculate evaporation rate and evapotranspiration are presented in Chapter 9. The subjects covered are the bulk aerodynamic equation, the Penman equation, evapotranspiration determination using the Bowen ratio method, a comparison of equations, eddy correlation, evaporation from a rough surface, calculated evapotranspiration from remotely sensed reflected solar radiation and surface temperatures with ground-station data, evaluation of evaporation using airborne radiometry and ground-based meteorological data, estimation of evaporation using thermal infrared satellite imagery, and evaporation from heated water bodies.

Two chapters are devoted to field instrumentation for measuring evaporation and evapotranspiration. In Chapter 10, evaporation pans are treated under the following topics: the Class A pan, the Young screened pan, the Colorado sunken pan, the sunken pan of the Bureau of Plant Industry, the GGI-1000 pan, the 20-m^2 basin, the Los Angeles County Flood Control District pans L and G, floating pans, the heat balance of the Class A evaporation pan, simultaneous recording of pan evaporation and rainfall, differences in evaporation pan data, evaporation from non-marine brines, the effect of salinity and ionic composition on evaporation, and pan-evaporation data as a practical tool for estimating potential evapotranspiration. In Chapter 11, lysimetry is treated under the following topics: lysimetry and evapotranspiration, hydraulic lysimeters, large weighing monolithic lysimeters, accuracy of lysimeters, accuracy of an hydraulic lysimeter, evaporation rate and lysimeters, direct evaluation of soil water flux after irrigation, and use of lysimeter to schedule irrigation automatically.

In the final chapter, evaporation reduction by various means is presented under the following subjects: effect of reflection of solar energy on evaporation of water, reduction of evaporation from water surfaces by the use of reflective surfaces, evaporation reduction by reduction of available solar energy, reduction of evaporation by the use of floating covers, evaporation reduction in stock tanks, reduction of evaporation by floating rubber or sponge tank covers, and effects of lily pads on evaporation.

Since this book is primarily a review of the literature, although some of my own work is cited, I am indebted to many authors who published as early as 1871 and as recently as 1990. I am also grateful to the U.S. National Weather Service, and particularly to Albert K. Showalter, who financially supported some of my early work on the evaporation of water. I am also grateful to the various organizations that provided the photographs reproduced in this book, and to the personnel of the U.S. Department of Agriculture Agricultural Research Service Water Conservation Laboratory in Phoenix, Arizona who were very cooperative and provided copies of many useful publications.

Frank E. Jones
32 Orchard Way South
Potomac, MD 20854

The Author

Frank E. Jones is currently an independent consultant. He received his Bachelor's degree in Physics from Waynesburg College, his Master's degree in Physics from the University of Maryland, and has also pursued doctoral studies in Meteorology at the University of Maryland. He served as a Physicist at the National Bureau of Standards (now the National Institute of Standards and Technology, NIST) in many areas including evaporation of water, humidity sensing, evapotranspiration, cloud physics, earthquake research, chemical engineering, processing of nuclear materials, mass, length, time, flow measurement, volume, and sound. Mr. Jones began work as an independent consultant upon his retirement from NIST in 1987. He is the author of more than 70 technical papers, one published book, and two other books that are now in production, and he also holds two patents. He is an Associate Editor of the National Council of Standards Laboratories Newsletter. He is a member of ASTM, the Instrument Society of America, the American Industrial Hygiene Association, the Institute for Nuclear Materials Management, and is associated with other technical societies from time to time as they are relevant to his interests.

Table of Contents

CHAPTER 1

Introduction

Water is the most abundant[1] and one of the most complex of known compounds on the surface of the earth.

The volume of the oceans is 1.32×10^{18} m^3, 3.49×10^{20} gal, or 1.07×10^{15} ac-ft (1 m^3 = 264.2 gal, 1 ac-ft = 3.259×10^5 gal = 1234 m^3).[2] The volume of the freshwater lakes is 1.25×10^{14} m^3, 3.30×10^{16} gal, or 1.01×10^{11} ac-ft. The volume of the saline lakes and inland seas in 1.04×10^{14} m^3, 2.75×10^{16} gal, or 8.43×10^{10} ac-ft. The volume of the rivers (average instantaneous volume) is 1.25×10^{12} m^3, 3.30×10^{14} gal, or 1.01×10^9 ac-ft. The volume of the soil moisture and vadose zone is 6.7×10^{13} m^3, 1.77×10^{16} gal, or 5.43×10^{10} ac-ft. The vadose zone is the unsaturated zone above the water table.

The annual precipitation on the ocean and land areas of the earth is 4.20×10^{14} m^3, 1.11×10^{17} gal, or 3.40×10^{11} ac-ft. The annual evaporation from ocean and land areas is the same as the annual precipitation.

Evaporation of water is a major factor in hydrologic systems. In general, an understanding of hydrologic systems requires an understanding of evaporation. There are many examples of the need for such understanding; several of these are mentioned here.

Less than 0.027% of the total amount of water distributed over the earth is fresh and immediately available.[3] The steadily increasing need for water requires an adequate knowledge of water management and the water balance. An important term in this balance is the evaporative loss of water from a free-water surface, from soil surfaces, or by evapotranspiration of crops.[4] Evapotranspiration can be defined as the loss of water from soil and plant surfaces.[5] The word evapotranspiration is often used when it is not possible to separate evaporation and plant transpiration. Consistent measurement of evaporation is crucial to utilization of evaporation data for water management.[6] A knowledge of the magnitude and variation of evaporative losses is required for the design and management of many hydrologic systems.[7]

The most economical means of providing additional usable water supplies is the conservation of water contained in existing storage facilities.[8] Evaporative losses from such storage facilities can be relatively great. For example, measurements have shown that 15.6 million ac-ft of water is lost from storage impoundments in the 17 western states of the United States each year.[9] This loss is equivalent to the disappearance of all the usable stored water in California (in 1965).

Since much of the available supply of water is stored in reservoirs, emphasis has been placed on preserving this water for future use. One approach to attempting to conserve water stored in reservoirs has been the application of a monolayer to the water surface to reduce evaporation.[10]

Evaporation is important in determining the water balance of watersheds, allowing prediction and estimation of runoff and groundwater recharge. Evaporation from soils is an important factor in managing both irrigated and dryland farming operations.[11]

Agricultural irrigation is the largest consumptive use of water in the United States. The total withdrawal of water for irrigation in the United States in 1980 was 170 million ac-ft, of which 40% was groundwater. This was 81% of all the withdrawn water used consumptively.[12] There are about 60 million irrigated acres in the United States and about 500 million irrigated acres in the rest of the world.[13] In Arizona and California, about 85% of the total water use is for crop irrigation. Most of the irrigation systems are surface or gravity systems, which typically have efficiencies of 60 to 70%.[14] Measurement of evapotranspiration is necessary to obtain engineering data on the time pattern of water use by crops and for irrigation criteria.[15] Estimates of evapotranspiration are used extensively in assessing the irrigation water-management efficiency of existing projects.[16] The meteorological variables which characterize the different seasons of the year produce large changes in the rate of evaporation from exposed surfaces after irrigation.[17]

A weighing lysimeter has been used as a feedback irrigation controller to measure crop evapotranspiration and simultaneously schedule irrigations for the lysimeter and three drip irrigation systems in a surrounding experimental field.[18] Pan evaporation measurements are useful in water use projections, such as forecasts of irrigation water demand for crops, estimates of losses from percolation ponds used for groundwater recharge, and the design of ponds for concentrating brines and effluents from waste disposal facilities.[19]

Quantitative forecasting (both hydrologic and meteorologic), the radar measurement of rainfall, the planning of water supplies for sprinkler irrigation, and many other related problems may require a knowledge of water lost by evaporation of water drops. These drops may be falling raindrops, spray from sprinklers or breaking waves, and raindrops falling on vegetation.[20] A model for accurate prediction and separation of the losses due to evaporation and wind drift under varying climatic conditions would be of considerable value to designers of sprinkler systems.[21]

Attempts have been made to stabilize aqueous fogs and atmospheric mists using monolayers.[22,23] Retardation of water fog evaporation could be used in diverse applications including blanketing of frost-threatened crops with a radiative barrier and military optical screening.[24]

The possibility of evaporating water at lower temperatures to reduce scaling and corrosion problems has been explored.[25] In problems of practical importance, such as liquid film cooling, drying of solids, and evaporation of water from large reservoirs, simultaneous heat and mass transfer between gas-liquid phases is present.[26] Heat exchange at the air-water interface is of increasing importance in the abatement of thermal pollution, design of cooling ponds, and in modeling temperature as an important element of water quality.[27]

Studies of the effect of salinity and ionic composition on evaporation can be applied in many engineering applications including water balance calculations for saline lakes, salt production ponds, and evaporation ponds used for disposal of saline effluents.[28]

The evaporation of water from vegetated surfaces is one of the less understood aspects of the hydrologic cycle.[29]

REFERENCES

1. Eisenberg, D., and W. Kauzmann. *The Structure and Properties of Water* (New York: Oxford University Press, 1969), p. v.
2. van der Leeden, F., F. L. Troise, and D. K. Todd. *The Water Encyclopedia* (Chelsea, MI: Lewis Publishers, Inc., 1990), 2nd ed., p. 58 (Source: Mace, U. S. Geological Survey, 1967).
3. Franks, F., Ed. *Water, A Comprehensive Treatise, Vol. 1,* (New York: Plenum Publishing Corporation, 1972), p. 2.
4. Bloeman, G. W. "A High-Accuracy Recording Pan-Evaporimeter and Some of its Possibilities," *J. Hydrology* 39:159—173 (1978).
5. Jackson, R. D., "Evaluating Evapotranspiration at Local and Regional Scales," *Proc. IEEE* 73:1086—1096 (1985).
6. Gunderson, L. H. "Accounting for Discrepancies in Pan Evaporation Calculations," *Water Resour. Bull.* 25:573—579 (1989).
7. Warnaka, K., and L. Pochop. "Analyses of Equations for Free Water Evaporation Estimates," *J. Geophys. Res.* 93:979—984 (1988).
8. Cooley, K. R. "Evaporation Suppression for Conserving Water Supplies," in *Proceedings of the Water Harvesting Symposium,* U.S. Department of Agriculture, Agricultural Research Service, February 1975, 192—200.
9. La Mer, V. K., and T. W. Healy. "Evaporation of Water: Its Retardation by Monolayers," *Science* 148:36—42 (1965).
10. Bartholic, J. F., J. R. Runkles, and E. B. Stenmark. "Effects of a Monolayer on Reservoir Temperature and Evaporation," *Water Resour. Res.* 3:173—179 (1967).
11. Idso, S. B., R. D. Jackson, and R. J. Reginato. "Estimating Evaporation: A Technique Adaptable to Remote Sensing," *Science* 189:991—992 (1975).
12. Bouwer, H. "Water Conservation in Agricultural and Natural Systems," in *Proceedings of the Conference on Water for the 21st Century* (Southern Methodist University), 7:414—423 (1984).

13. Bouwer, H. "Effect of Irrigated Agriculture on Groundwater," *J. Irrig. Drain. Eng.* 113:4—15 (1987).
14. Bouwer, H. "Water Conservation," *Agric. Water Manage.* 14:233—241 (1988).
15. Tanner, C. B. "Measurement of Evapotranspiration," in *Irrigation of Agricultural Lands,* American Society of Agronomy Monograph No. 11, pp. 534—574.
16. Jensen, M. E. "Empirical Methods of Estimating or Predicting Evapotranspiration Using Radiation," in *Conference Proceedings: Evapotranspiration and its Role in Water Resources Management* (St. Joseph, MI: American Society of Agricultural Engineers, 1966), p. 49.
17. Kimball, B. A., and R. D. Jackson. "Seasonal Effects on Soil Drying after Irrigation," *Hydrology and Water Resources of Arizona and the Southwest Proceedings* (Arizona Section, American Water Research Association; and Hydrology Section, Arizona Academy of Science) 1:85—88 (1971).
18. Phene, C. J., R. L. McCormick, K. R. Davis, J. D. Pierro, and D. W. Meek. "A Lysimeter Feedback Irrigation Controller System for Evapotranspiration Measurements and Real Time Irrigation Scheduling," *Trans. ASAE* 32:477—484 (1989).
19. Goodridge, J. R. "Evaporation from Water Surfaces in California," *Calif. Dep. Water Resour. Bull.* 73—79 (1979).
20. Showalter, A. K. "Evaporative Capacity of Unsaturated Air," *Water Resour. Res.* 7:688—691 (1971).
21. Kincaid, D. C., and T. S. Longley. "A Water Droplet Evaporation and Temperature Model," *Trans. ASAE* 32:457—463 (1989).
22. Derjaguin, B. V., V. A. Fedoseyev, and L. A. Rosenzweig. "Investigation of the Adsorption of Cetyl Alcohol Vapor and the Effect of This Phenomenon on the Evaporation of Water Drops," *J. Colloid Interface Sci.* 22:45—50 (1966).
23. May, K. R. "Comments on 'Retardation of Water Drop Evaporation with Monomolecular Films'," *J. Atmos. Sci.* 29:784—785 (1972).
24. Carlon, H. R., and R. E. Shafer. "Optical Properties (0.63— 13 μm) of Water Fogs Stabilized Against Evaporation by Long-Chain Alcohol Coatings," *J. Colloid Interface Sci.* 82:203—207 (1981).
25. Kingdon, K. H. "Enhancement of the Evaporation of Water by Foreign Molecules Adsorbed on the Surface," *J. Phys. Chem.* 67:2732—2737 (1963).
26. Massaldi, H. A., J. C. Gottifredi, and J. J. Ronco. "Effect of Interfacial Waves on Mass Transfer During Evaporation of Water From a Free Surface," *Lat. Am. J. Chem. Eng. Appl. Chem.* 6:161—170 (1976).
27. Chattree, M., and S. Sengupta. "Heat Transfer and Evaporation From Heated Water Bodies," *J. Heat Transfer* 107:779—787 (1985).
28. Salhotra, A. M., E. Eric Adams, and D. R. F. Harleman. "Effect of Salinity and Ionic Composition on Evaporation: Analyses of Dead Sea Evaporation Pans," *Water Resour. Res.* 21:1336—1344 (1985).
29. Reginato, R. J., R. D. Jackson, and P. J. Pinter, Jr. "Evapotranspiration Calculated from Remote Multispectral and Ground Station Meteorological Data," *Remote Sensing Environ.* 18:75—89 (1985).

CHAPTER 2

Transport of Water Across the Interfacial Region

INTRODUCTION

A study of the evaporation of water logically starts at the water-gas interface. In this chapter, the investigation of the transport of water across the interfacial region begins with the kinetic theory of gases.

KINETIC THEORY OF GASES

From the kinetic theory of gases:[1]

$$\Phi_0 = {}^1/_4\, n\bar{c} \tag{1}$$

where ϕ_0 = the number of molecules striking unit area of a *surface in a gas* per unit time
n = the number of molecules of gas per unit volume
$\bar{c}$ = the mean molecular speed

The mean molecular speed can be computed from the Maxwell-Boltzmann probability distribution function in the form:[2]

$$P(c)\, dc = 4\pi\sqrt{(m/2\pi kT)^3}\; e^{-mc^2/2kT} c^2\, dc \tag{2}$$

where $P(c)\, dc$ = the probability that the molecule will have a speed, c, in the speed interval dc
m = the mass of the molecule
k = Boltzmann's constant,[3] $(1.3806513 \pm 0.0000025) \times 10^{-16}$ joule/K
T = the absolute temperature in °K

$$\bar{c} = \int_0^\infty c\, P(c)\, dc = 4\pi\sqrt{(m/2\pi kT)^3} \int_0^\infty c^3\, e^{(-mc^2/2kT)}\, dc \tag{3}$$

$$\int_0^\infty c^3\, e^{(-mc^2/2kT)}\, dc = {}^1/_2\, \Gamma(2)\, (2kT/m)^2 = {}^1/_2\, (2kT/m)^2 \tag{4}$$

where Γ = the gamma function

Therefore:

$$\bar{c} = \sqrt{(8kT/\pi m)} \tag{5}$$

The equation of state for an ideal gas:

$$\bar{p} = nkT \tag{6}$$

where $\bar{p}$ = the mean pressure, yields:

$$n = \bar{p}/kT \tag{7}$$

Therefore:

$$\phi_0 = {}^1/_4\, n\bar{c} = {}^1/_4\, (\bar{p}/kT)\sqrt{(8kT/\pi m)} = \bar{p}/\sqrt{(2\pi mkT)} \tag{8}$$

Before attempting to apply Equation 8 to the evaporation of water from (or condensation on) a liquid water surface, the *assumptions* on which the equation is based should be stated. They are

1. The four fundamental assumptions of the kinetic theory of gases as stated by Knudsen:[4] (a) "Any gas consists of separate particles called molecules. In a pure gas they are alike." (b) "The molecules move about in all directions." (c) "The pressure caused by the movement of the molecules is the only one existing in the gas, when it is in the ideal state." (d) "The molecules are not infinitely small. Thus, they collide with one another."
2. The speed of the gas (water vapor) molecule is distributed as in Equation 2.
3. As a first approximation the gas (water vapor) behaves as an ideal gas.

The applicability of Equation 8 depends on the validity of these assumptions.

The mass flux, J_m (g/cm^2·sec), is defined by:

$$J_m \equiv m\Phi_0 = \bar{p}m/\sqrt{(2\pi mkT)} = \bar{p}\sqrt{(m/2\pi kT)} \tag{9}$$

and the molar flux, J_M (mol/cm^2·sec), is defined by:

$$J_M \equiv m\Phi_0/M = J_m/M = \bar{p}\sqrt{(m/M^2k)}/\sqrt{(2\pi T)} = \bar{p}/\sqrt{(2\pi MRT)} \tag{10}$$

where M = the molecular weight (for water, $M = 18.0152$ g/mol)
R = the universal gas constant, $(8.314471 \pm 0.000014) \times 10^7$ joules/K·mol,[3] $R \equiv N_a k = Mk/m$, and N_a is Avogadro's number, $(6.0221367 \pm 0.0000036) \times 10^{23}$/mol^{-1}

(See Reference 3.)

It must be emphasized that the kinetic theory of gases deals with molecules striking an idealized surface in a gas, not a physical interface such as an air-water interface. In attempting to apply the kinetic theory result to the water evaporation system, one goes from the idealized to the physical system and eventually to experimental systems which in some respects do not satisfactorily represent either the idealized or the pure physical system. One must therefore be aware that the kinetic theory result might not be applicable without adjustment. In what follows, the kinetic theory result will be considered in connection with the water vapor-water, air-water, or vacuum-water interface, keeping these remarks in mind.

EVAPORATION AND CONDENSATION OF WATER

The *maximum possible* number of molecules, ϕ_s, (that is, the kinetic theory result of Equation 8) leaving unit area of plane water surface per unit time when the water surface is in *equilibrium* with saturated water vapor is equal to the number impinging on unit area per unit time from the saturated vapor:[5]

$$\Phi_s = e_s/\sqrt{(2\pi m k T_s)} \qquad (11)$$

where e_s = the saturation vapor pressure over a plane water surface at the temperature, T_s, of the water surface

The corresponding mass flux is

$$J_m = e_s\sqrt{(m/2\pi k T_s)} \qquad (12)$$

and the corresponding molar flux is

$$J_M = e_s/\sqrt{(2\pi M R T_s)} \qquad (13)$$

where the molecular weight M of water is, again, 18.0152 g/mol

For saturated water vapor at 20°C, Equation 11 yields 8.5×10^{21} molecules impinging on 1 cm^2 of surface per second; Equation 12 yields a mass flux of 0.25 g/cm^2·sec; and Equation 13 yields a molar flux of 0.014 mol/cm^2·sec. At these theoretical rates, assuming a constant surface temperature, the water in Lake Mead in Nevada would completely evaporate in about 6 hr! Obviously,

Figure 1. Aerial view of Hoover Dam and Lake Mead showing a nearly full reservoir. (From Bureau of Reclamation, U.S. Department of Interior.)

these theoretical rates are *much higher* than those encountered in the field. In what follows, material will be presented which will account for much of the discrepancy. Lake Mead is shown in Figure 1.

Introducing the concept, attributed to Knudsen,[5] that only a fraction, E (the evaporation coefficient), of the molecules crossing a plane adjacent to the liquid surface comes from the liquid, Equation 12 becomes:

$$J_+ = E\, e_s\sqrt{(m/2\pi kT_s)} \tag{14}$$

where J_+ = the mass flux normal to the plane and *away* from the liquid surface

If the water vapor pressure in the space adjacent to the liquid surface is maintained at some value, e:

$$J_- = C\, e\sqrt{(m/2\pi kT_v)} \tag{15}$$

where J_- = the mass flux normal to the plane and *toward* the liquid surface
C = the fraction (the condensation coefficient) of impinging molecules which condense on the liquid surface
T_v = the temperature of the vapor

PRINCIPLE OF DETAILED BALANCING

At *equilibrium,* the relationship between E and C is found by the direct application of the principle of detailed balancing: "when a system is in equilibrium any single process balances the reverse process."[6] Since the evaporation process is the reverse of the condensation process, and assuming that a state of equilibrium exists, the principle of detailed balancing yields:[7]

$$E = C \tag{16}$$

By definition, $T_v = T_s$; therefore, the net mass flux, which is called the evaporation rate away from the liquid surface is given by:

$$J_+ - J_- = E\sqrt{(m/2\pi kT_s)}\,(e_s - e) = E\sqrt{(M/2\pi RT_s)}\,(e_s - e) \tag{17}$$

An equation of the form:

$$(J_+ - J_-)/m = \Phi_o = E\sqrt{(1/2\pi mkT_s)}\,(e_s - e) \tag{18}$$

is called the *Hertz-Knudsen* equation.[8]

In the preceding development of the Hertz-Knudsen equation, the evaporation coefficient, *E,* has been treated as a coefficient, and there has been no attempt at a physical interpretation. The water surface (or gas-water interface) was treated as an idealized surface with a determinable temperature, T_s, with no other limitation on the transport of water vapor than that imposed by difference in water vapor pressure across the surface. Also, there was no consideration of the sources of the latent heat of evaporation. Experimentally determined values of the evaporation coefficient would therefore be expected to reflect the nonideality of the experimental situation. In what follows, a somewhat different approach will be taken leading to physical interpretations of the coefficient.

ABSOLUTE REACTION RATES

An equation of the form of Equation 18 has been derived[9] from the theory of absolute reaction rates:[10]

$$v = \kappa\, e/\sqrt{(2\pi mkT)} \qquad (19)$$

where v = the rate of evaporation (molecules/cm²·sec) of a liquid
e = the vapor pressure
κ = the transmission coefficient (the fraction of the number of surface water molecules which proceed directly to the vapor phase without returning to the water surface)

The transmission coefficient is also called the condensation coefficient and is equal (again, through the application of the principle of detailed balancing) to the fraction of the number of surface vapor phase molecules which proceed into the water surface without returning to the vapor phase.

$$\kappa = F_e/F_i \qquad (20)$$

where F_e = the molecular partition function per unit volume for the surface molecules
F_i = the internal part of the molecular partition function per unit volume for the molecules in the vapor phase

The partition function per unit volume, F, of a molecule is the probability of occurrence of the molecule in the specified volume and is equal to the sum of the exp($-\epsilon$/kT)-terms for all forms of energy, ϵ, i.e., translational, vibrational, rotational, electronic, and nuclear, possessed by the molecule. Each term is weighted according to the degeneracy of the particular energy level.[11] Thermodynamic quantities such as equilibrium constants, free energies, and entropies can be obtained from partition functions.

If it is assumed that the different forms of energy distribution are independent of each other, the complete partition function may be taken to be equal to the product of the functions for the separate energies. For the water molecule, then:

$$F = F_{tr.}\cdot F_{vib.}\cdot F_{rot.} \qquad (21)$$

The subscripts indicate the separate energies: tr., translational; vib., vibrational; and rot., rotational. The translational motion away from the surface was omitted in the derivation of Equation 20.[9] The vibrational partition functions are eliminated on the assumption that partition function of internal vibrations is unaltered by the transition from the liquid to the vapor phase.[12,13] Therefore, κ is the ratio of rotational partition functions for surface molecules to those for molecules in the vapor phase.

Kincaid and Eyring[14] evaluated this ratio as the "free angle ratio," δ. Mortensen and Eyring[9] tabulated a comparison of free angle ratios, i.e., κ.

t, °C	δ_1(K&E)	δ_2(Tr)	δ_3(H)	δ_4(P)	Observed κ
0	0.022	0.11	0.099	0.17	0.036 (15°C)
100	0.17	0.11	0.067	0.17	0.02 (100°C)

K&E refers to δ calculated by the method of Kincaid and Eyring; Tr refers to δ obtained from the reduced temperature (the ratio of the temperature of the liquid, T, to its critical temperature, T_c); H refers to δ calculated from Hildebrand's rule[15] that the entropies of vaporization should be compared at equal vapor concentrations; and P refers to δ calculated from Pitzer's rule[16] that entropies of vaporization should be compared at equal vapor-to-liquid volume ratios.

The value 0.036 for κ at 15°C was taken from Alty and Mackay[17] and the value at 0.02 at 100°C was taken from Prüger.[18] Mortensen and Eyring[9] observed from their comparison that polar molecules (water, etc.) have small free angle ratios and small condensation coefficients indicating a small rotational partition function on the surface compared to the rotational partition function in the vapor phase; that this is a consequence of orientation of the molecules due to the strong electrostatic forces acting on them; and that, since liquid molecules possess a cooperative structure which is quite different from that of a solid or gas, it is not surprising that a molecule the rotation of which cannot pass adiabatically into the liquid structure should be rejected at the water surface.

The theory of absolute reaction rates thus provides physical justification for the existence of the evaporation coefficient for water and permits the calculation of values which are in reasonable agreement with experimentally determined values.

CONCEPT OF DIFFUSIONAL RESISTANCE

In the literature on the experimental determination of evaporation and condensation coefficients, the concept of the diffusional resistance to the transport of water is encountered. It is appropriate to introduce the concept here. It serves to outline the transport of water from the interior of the liquid, through the interfacial region, and through the gas phase.

The concept of the series diffusional resistance representation was apparently propounded by Lewis and Whitman[19] in connection with gas absorption, and was treated in detail by Langmuir and Langmuir[20] and Davies and Rideal.[21] Davies and Rideal treated the transport of material across any plane surface by using a linear approximation to a diffusion law in one dimension:

$$dj_M/dt = AD\,\Delta N/\Delta x \tag{22}$$

where j_M = the number of moles transported
t = the time
A = the area across which the material is transported
D = a diffusion coefficient
$\Delta N/\Delta x$ = the concentration gradient

A permeability coefficient, Λ(cm/sec), the reciprocal of a diffusional resistance, R(sec/cm), is defined by:

$$\Lambda \equiv 1/R = D/\Delta x \tag{23}$$

and, from kinetic theory, the diffusional resistance, R_I, for a clean liquid surface is given by:

$$R_I = (1/E)\sqrt{(2\pi m/kT}} \tag{24}$$

A molecule crossing the liquid-gas interface, on the basis of the diffusional concept, encounters a total resistance, R, the sum of three diffusional resistances in series:

$$R = R_G + R_I + R_L \tag{25}$$

where R_G = the diffusional resistance in the gas phase
R_i = the diffusional resistance of the monomolecular region constituting the liquid-gas interface
R_l = the diffusional resistance of the liquid beneath the interfacial region

The diffusional resistance concept provides a convenient separation of the study of the evaporation of water into investigations of three regions: (1) the gas phase, (2) the liquid-gas interfacial region, and (3) the liquid phase.

HEAT AND MASS TRANSFER

Although some of the concepts of the heat and mass transfer approach are implicit in several areas in this book, the subject will be dealt with explicitly here in the interest of completeness.

Both energy (heat) and mass are transferred when a liquid evaporates. The rate at which the liquid evaporates depends on the rate at which energy can be supplied to the interfacial region. The coupled transport of heat and mass can be described through the introduction of heat-transfer and mass-transfer coefficients.

In what follows, the heat- and mass-transfer concepts will be outlined and applied to evaporation. The approach taken by Bird, Stewart, and Lightfoot[22] will be followed closely. Heat- and mass-transfer coefficients will be defined, an average mass-transfer coefficient for small transfer rates for the case of a gas mixture flowing along a mass-transfer surface will be defined. The dependence of the transfer coefficients on mass-transfer rates for high mass-transfer rates will be indicated. Results of the use of film theory for the prediction of the variation of the heat-transfer and mass-transfer coefficients with mass-transfer rate will be presented and the use of one of the resulting equations in an example for estimating the local evaporation of a liquid from a wetted porous slab submerged in a tangentially flowing stream of noncondensable gas will be indicated. The solutions of the Prandtl boundary layer equations to investigate the simultaneous transfer of momentum, heat, and mass across the boundary of the flowing liquid in terms of the fluxes at the boundary will be presented. The local transfer coefficients obtained directly from the fluxes at the boundary will be presented. The results of boundary layer theory will be restated in terms of the quantities that appeared in the film theory, and the use of the boundary layer results to estimate local evaporation in the previously mentioned example will be indicated. Finally, turbulent or eddy transfer will be discussed.

The heat-transfer coefficient, Θ_h (cal/cm^2·sec·deg), is defined by:

$$Q = \Theta_h A \, \Delta T \tag{26}$$

where Q (cal/sec) = the heat transferred across a characteristic interfacial area A (cm^2)
ΔT = the characteristic temperature difference

A mass-transfer coefficient for a liquid, Θ_ℓ (cm/sec), is defined by:

$$(J_M)_{AO} = \Theta_\ell \, \Delta N_A \tag{27}$$

where $(J_M)_{AO}$ (mol/cm^2·sec) = the rate of transfer of the liquid (molar flux)
ΔN_A (mol/cm^3) = molar concentration difference

Similarly, for a gas, Θ_g (mol/sec·dyne) is defined by:

$$(J_M)_{AO} = \Theta_g \, \Delta p_A \tag{28}$$

where Δp_A (dyne/cm^2) = partial pressure difference

For the case of a flowing stream of gases A and Z along a mass-transfer surface, for small transfer rates, an average mass-transfer coefficient, Θ_x (mol/cm²·sec), is defined by:

$$W_A = \chi_{AO}(W_A + W_Z) = \Theta_x S \, \Delta\chi_A \quad (29)$$

where W_A and W_Z are the molar rates of addition of A and Z to the stream over the entire surface

χ_{AO} = $1 - \chi_{ZO}$ is some characteristic mole fraction of A next to the interface

S = a surface of finite area

$\Delta\chi_A$ = a characteristic composition difference

The left side of Equation 29 is the rate of diffusion of A into the fluid at the interface and corresponds to Q in Equation 26; $\Delta\chi_A$ corresponds to ΔT and Θ_x corresponds to Θ_h of Equation 26. For high mass-transfer rates, the transfer coefficients depend on the mass-transfer rates; this dependence is indicated by the superscript (*), e.g., Θ_h^* and Θ_x^*.

Using the film theory[23-25] to predict the variation of heat-transfer and mass-transfer coefficients with mass-transfer rate, the following equations result:

$$1 + (T_O - T_\infty)\,[(J_M)_{AO}C_{pA} + (J_M)_{ZO}C_{pZ}]/Q_O = \exp\{[(J_M)_{AO}C_{pA} + (J_M)_{ZO}C_{pZ}]/\Theta_{h,loc}\} \quad (30)$$

$$1 + (1 - \chi_{A\infty}/\chi_{AO})\chi_{AO}/\left\{\{(J_M)_{AO}/[(J_M)_{AO} + (J_M)_{ZO}]\} - \chi_{AO}\right\} = \exp\{[(J_M)_{AO} + (J_M)_{ZO})]/\Theta_{x,loc}\} \quad (31)$$

where T_0 = the temperature of the interface

T_∞ = the "free-stream" temperature

$(J_M)_{AO}$ and $(J_M)_{ZO}$ are the constant fluxes of A and Z at the interface

C_{pA} and C_{pZ} are the heat capacities (cal/mol·deg) of A and Z

Q_O = the heat flux (cal/cm²·sec) at the interface

$\Theta_{h,loc}$ = the local heat-transfer coefficient

χ_{AO} = the mole fraction of A at the interface

$\chi_{A\infty}$ = the "free-stream" mole fraction of A

$\Theta_{x,loc}$ = the local mass-transfer coefficient

Taking the natural logarithm of both sides of Equation 31:

$$\ln R = [J_M)_{AO} + (J_M)_{ZO}]/\Theta_{x,loc} \quad (32)$$

where R is the left side of Equation 31

Rearranging Equation 32:

$$(J_M)_{AO} = \Theta_{x,loc} \ln R - (J_M)_{ZO} \tag{33}$$

Bird et al.[22] used an equation of the form of Equation 33 in an example to estimate the local rate of evaporation of a liquid A from a wetted porous slab submerged in a tangentially flowing stream of a noncondensable gas Z and arrived at a lower evaporation rate, they said, than is obtained by boundary-layer theory.

Solution of the Prandtl Boundary-Layer Equations

Bird et al.[22] solved the Prandtl boundary-layer equations exactly to determine what happens when heat, mass, and momentum are simultaneously transferred across the boundary of a flowing liquid. The system considered consisted of a thin semi-infinite plate of volatile solid A subliming, under steady conditions, into an unbounded gaseous stream of A and Z (species Z being present in the gas phase only). The following were assumed: known uniform temperature and gas composition along the plate surface; no chemical reactions and no external forces other than gravity; no viscous dissipation, heats of mixing, or emission or absorption of radiant energy in the gaseous boundary layer; constant density, ρ; viscosity, μ; heat capacity, C_P; thermal conductivity, κ; molar concentration, χ; and mass diffusivity, D_{AZ}.

With these assumptions, the Prandtl boudary layer equations reduced to:
(continuity)

$$(\partial v_x/\partial x) + (\partial v_y/\partial x) = 0 \tag{34}$$

(motion)

$$v_x(\partial v_x/\partial x) + v_y(\partial v_y/\partial y) = \nu(\partial^2 v_x/\partial y^2) \tag{35}$$

(energy)

$$v_x (\partial T/\partial x) + v_y(\partial T/\partial y) = D_T(\partial^2 T/\partial y^2) \tag{36}$$

(continuity of A)

$$v_x(\partial \chi_A/\partial x) + v_y(\partial \chi_A/\partial y) = D_{AZ}(\partial^2 \chi_A/\partial y^2) \tag{37}$$

where v = the components of the stream velocity
x and y are the two-dimensional Cartesian coordinates
ν = the kinematic viscosity

D_T = the thermal diffusivity

The boundary conditions are x less than or equal to 0 or $y = \infty$, $v_x = v_\infty$, $T = T_\infty$, and $\chi_A = \chi_{A\infty}$; at $y = 0$, $v_x = 0$, $T = T_0$, $\chi_A = \chi_{AO}$, and $(J_M)_Z = 0$. The fluxes of momentum, energy, and mass at the boundary are given by the resulting dimensionless expressions:

$$(\tau_{yx}|y = 0)/\rho v_\infty(0 - v_\infty) = (\Pi'(0,1,K)(\sqrt{\nu/v_\infty x}))/2 \quad (38)$$

$$(Q_y|y = 0)/\rho C_p v_\infty(T_0 - T_\infty) = (\Pi'(0,Y_T,K)(\sqrt{(\nu/v_\infty x)}))/2Y_T \quad (39)$$

$$(J^*_{Ay}|y = 0)/N_M v_\infty(\chi_{AO} - \chi_{A\infty}) = (\Pi'(0,Y_{AZ},K)(\sqrt{(\nu/v_\infty x)}))/2Y_{AZ} \quad (40)$$

where $\tau_{yx}|\, y = 0$ is the x-directed shear stress on the boundary resulting from viscous forces, evaluated at the boundary
$Q_y|\, y = 0$ is the energy flux at the boundary
$J^*_{Ay}|\, y = 0$ is the molar diffusion flux of species A at the boundary relative to the molar average velocity

$\Pi'(0,1,K)$ is the dimensionless gradient of velocity at the boundary; K is the dimensionless mass-transfer rate defined by $K \equiv (2v_{yo}/v_\infty)(v_\infty x/\nu)^{1/2}$ (where v_{yo} is the fluid velocity in the y-direction at the wall; $\Pi'(0,Y_T,K)$ is the dimensionless gradient of the temperature at the boundary; $Y_T = \nu/D_T \equiv Pr$, the Prandtl number; $\Pi'(0,Y_{AZ},K)$ is the dimensionless gradient of concentration at the boundary; and $Y_{AZ} = \nu/D_{AZ} \equiv Sc$, the Schmidt number)

The local transfer coefficients for simultaneous momentum, heat, and mass transfer are obtained directly from Equations 38—40 through the following equations:

$$(f^*_{loc}/2) = \Pi'(0,1,K)R' \quad (41)$$

$$(\Theta^*_{loc}/\rho\hat{C}_p v_\infty) = \Pi'(0,Pr,K)R'/Pr \quad (42)$$

$$(\Theta^*_{X,loc}/cv_\infty) = \Pi'(0,Sc,K)R'/Sc \quad (43)$$

where $R' \equiv (v_\infty x/\nu)^{-1/2}/2$
$\hat{C}_p$ = heat capacity per unit mass (cal/mol·deg)
c = molar concentration (mol/cm³)

The results of boundary layer theory, Equations 38—40, were restated in terms of the quantities that appeared in the film theory:

J_{AZ} (dimensionless molar flux ratio) =

$$((J_M)_{AO} + (J_M)_{ZO})/\Theta^{*}_{X,loc} = \\ (\chi_{AO} - \chi_{A\infty})/\left\{\{(J_M)_{AO}/[(J_M)_{AO} + (J_M)_{ZO}]\} - \chi_{AO}\right\} \tag{44}$$

D_{AZ} (dimensionless molar diffusion rate ratio) =

$$[(J_M)_{AO} + (J_M)_{ZO}]/\Theta_{X,loc} \tag{45}$$

When all physical properties are constant in the mixture and equal for species A and Z, Equation 44 becomes:

$$J_{AZ} = cv_{yo}/\Theta^{*}_{X} \tag{46}$$

Inserting the boundary layer solution for Θ^{*}_{X}:

$$J_{AZ} = 2\ Sc\ K\ cv_{yo}/2\ (v_{yo}/v_{\infty})cv_{\infty}\Pi'(0,Sc,K) = K\ Sc/\Pi'(0,Sc,K) \tag{47}$$

$$D_{AZ} = K\ Sc/\Pi'(0,Sc,0) \tag{48}$$

Setting $(J_M)_{ZO} = 0$, since species Z is noncondensable, Equation 45 becomes:

$$(J_M)_{AO} = \Theta_{X,loc}D_{AZ} \tag{49}$$

FLOW IN A FULLY DEVELOPED TURBULENT BOUNDARY LAYER

The boundary-layer theory would be more accurate than the film theory solution for the evaporation of water into a laminar boundary layer, although its accuracy depends on the validity of the many assumptions made in the solution of the Prandtl boundary-layer equations. However, in the atmosphere, the mass-transfer surface may be considered to be hydrodynamically rough and the air flow just above the surface layer may be considered to be turbulent.[26] Therefore, the boundary layer of interest for evaporation of water is a turbulent boundary layer. Flow in a fully developed turbulent layer is discussed in detail in several sources including Sutton.[27] It will not be discussed here in detail; instead some items of particular interest to the present discussion will be presented.

Sutton gives the following description of several regions of flow in a fully developed turbulent boundary layer of a smooth (in the ordinary sense of the word) boundary:

1. Immediately adjacent to the surface, a very thin laminar sublayer, within which vertical eddy motions are practically non-existent. Inside this layer the velocity gradient $d\bar{u}/dz$ attains very high values, and the shearing stress is effectively that caused by viscosity alone.
2. Above this, the turbulent boundary layer proper, characterized by strong vertical motions and a small gradient of mean velocity. In this layer the Reynolds stress is at least as great as the viscous stress and may be much greater.
3. Above the boundary layers, the free stream, in which the viscous stresses are negligible.

Reynolds stresses are terms of the form $-\rho\, \overline{u_i' u_j'}$, where u' is a fluctuation of a component of the wind, $i,j = 1,2,3$.

In the development of a theory of turbulence, the transfer of momentum, heat, and mass is expressed in terms of virtual coefficients of viscosity, conductivity, and diffusivity called "Austausch coefficients" or "exchange coefficients."[28]

One of the statements of the exchange coefficient hypothesis is

$$E = -\rho K_w\, \partial\bar{q}/\partial z \tag{50}$$

where E (g/cm²·sec) is the turbulent or eddy flux of water vapor
ρ (g/cm³) is the mean density
K_w is the turbulent or eddy diffusivity of water
$\partial\bar{q}/\partial z$ (cm⁻¹) is the vertical gradient of the mean specific humidity

In principle, then, the evaporation rate, E, could be determined by a determination of $\partial\bar{q}/\partial z$, assuming ρ and K_w to be known constants. This approach is discussed in Chapter 9.

EVAPORATION BY SPURTS

Kingdon[29] explored experimentally the concept that adsorbed foreign molecules on the surface of water, which are capable of forming weaker hydrogen bonds with water molecules than occur in normal water, will increase the rate of evaporation of water. Kingdon's premise was that the evaporation of water molecules is controlled by the hydrogen bonds between them, and that it is expected that if foreign molecules which can form weaker hydrogen bonds with water molecules are added to the surface, the rate of evaporation of molecules will increase. The overall motivation of the experimental work was to explore the possibility of evaporating water at lower temperature so as to reduce scaling and corrosion problems.

Water was evaporated from a stainless steel dish, 8.4 cm (3.3 in.) in diameter, usually heated to 63°C. A mass spectrometer was used to measure the water vapor content close to the evaporating water surface. The mass spectrometer was thus an indicator of the rate of evaporation with rapid response and capability of detecting a 1% change in the rate under some experimental conditions. Spurts of evaporation were observed with atmospheres of air, nitrogen, carbon dioxide, methane, and butane. Of the gases tested, helium and hydrogen were the only two which showed smooth evaporation. The spurts were more frequent at higher gas flows across the water surface.

Bulk distillation of water was carried out in atmospheres of 11 different gases: the largest rates occurring in argon, oxygen, and nitrogen, and the smallest in hydrogen and helium. For evaporation into streams of various gases, butane gave the largest rate and helium the smallest. These results were systematized by considering the effects of intermolecular forces as represented by the diffusion constants, which determine both the diffusion in the gas phase and the spurt-producing interaction of the foreign gas with the water surface.

EVAPORATION OF WATER IN ELECTRIC FIELDS

Carlon[30] discovered and studied the acceleration of the evaporation of water from moistened, porous substrates in electric fields.

Thin substrates of cotton cloth or paper toweling were wetted with water and placed on two flat, parallel electrodes to which high voltages (up to 9000 volts, V, dc) were applied. The electric current flowing between the electrodes was monitored for various voltage settings.

The rate of evaporation increased nearly 10-fold compared to the evaporation rate in still air with no voltage applied, for field strengths greater than 6000 V/cm. The rate of water evaporation was found to be proportional to the electric field strength between the electrodes. The magnitude of the electric currents flowing between the electrodes increased by a factor of 10^3 to 10^4 compared to currents for dry substrates or bare electrodes. Drying was uniform across the substate surfaces.

When the electrode on which the wetted substrate lay was positively charged, the system worked best possibly indicating that water ion species such as $H^+(H_2O)_n$ were the charge carriers and were repelled away from the like-charged electrode in larger numbers than were negative water ions such as $OH^-(H_2O)_n$. Preliminary results indicated that water and water solutions exhibited the behavior described in the paper to the greatest extent, but that oils and other organic liquids also exhibited this behavior to varying degrees.

Carlon believed that this phenomenon had not been previously discovered and reported.

PREPARATION OF PURE WATER

The procedures used in the work of Gittens[31] to prepare pure water is of sufficient interest to warrant quotation

> Water was purified by a primary distillation in an all-glass apparatus followed by a second distillation in a well-seasoned Pyrex™* apparatus from neutral permanganate, against a nitrogen backflush, through a 3-foot fractionating column filled with stainless steel mesh and fitted with a double splash head. Water was collected and stored in a receiver, heated to a temperature close to the experimental temperature, and kept under nitrogen. Samples were removed under nitrogen pressure. White spot nitrogen was used after washing with phosphoric acid and water....water had a specific conductivity of less than 10^{-9}/ohm·m. For each batch of water, surface tension values were checked at 25°C before and after a series of measurements, all results for the batch were rejected if significant deviation had occurred. The cleanliness of the system was demonstrated by the observation that constant and reproducible values of surface tension were maintained over periods of weeks.

Drost-Hansen[32] discussed the preparation of very pure water for surface tension and interfacial tension measurements and concluded that water for surface studies must be obtained by distillation. The methods of Franks,[33] Taylor,[34] Butler,[35] and Eigen and De Maeyer[36] were mentioned.

MacIntyre[37] described a continuous still for producing high purity water. He reported that after two years of operation, the product collected at 90°C had a conductivity of less than 0.10 micromho/cm compared to the theoretical limiting value of 0.055 micromho/cm for pure water.

Smith[38] discussed problems in the production and handling of ultrapure water, water having a total solids count of less than 0.1 ppm and a conductivity of 1 micromho/cm at 25°C.

Simon and Calmon[39] reviewed the subject of water purity, of both ionic content and insoluble content; discussed some of the problems encountered in the production of ultrapure water; and illustrated equipment for its production and described facilities for washing integrated transistor circuits. The average values for the analytical properties in tentative specification formulated by industries in the semiconductor and integrated circuit field quoted are "conductivity", 16 megohm·cm; total electrolytes, 35 ppb as NaCl; particulate count 130/ml; nominal maximum size, 0.5 μm (2×10^{-5} in.); organics, by CO_2 formation, 1.0 ppm; dissolved gases, 200 ppm; and living organisms, 9/ml.

---

*Registered trademark of E. I. du Pont de Nemours and Company, Inc., Wilmington, Delaware.

The following ideas for future developments were presented for consideration:[39] "(1) Upgrading ultra pure water purity by zone freezing. (2) Transportation of ultra pure water as steam without wall contact followed by condensation at point of use. (3) Storage of ultra pure water as blocks of ice frozen to noncontaminating cold surfaces. Principles of zone freezing could be used here to discard the contaminated portions and use only the purest portions of the ice block."

A commercial available system for producing water of the "highest purity" is described in a recent paper.[40] The system is classified Type I RGW by the American Society for Testing and Materials. RGW is the abbreviation for reagent grade water and Type I indicates highest purity. The system is a wall-mounted disposable cartridge system using carbon absorption for the removal of chlorine and organics, serial mixed-bed ionization to remove ionized contaminants, and 0.2 μm (8×10^{-6} in.) final filtration for removal of particulates and bacteria. There is no storage tank since 2 L/min of water is produced on demand. A resistivity meter indicates the purity of water as it is being produced at the point of use. Type I RGW is meant to be produced at the point of use. The water rapidly deteriorates due to absorption of CO_2 and other airborne contaminants; it is also very aggressive and can attack a container, leaching out additional contaminants.

The design and operation of a pharmaceutical ultrapure water system were described by Nykanen and Cutler.[41] The sequence of processes in the system are softening; coarse filtration; carbon absorption; chemical conditioning; reverse osmosis; vacuum degasification; short-term permeate storage; recirculation through ultraviolet irradiation and mixed-bed deionization; and six-effect distillation.

In storage, the water was maintained hot at 70°C ± 2°C. The resistivity of the delivered water was greater than 0.5 megohm·cm.

REFERENCES

1. Knudsen, M., *The Kinetic Theory of Gases* (New York: John Wiley & Sons, Inc., 1950), 3rd ed., p. 2.
2. Kittel, C. *Elementary Statistical Physics* (New York: John Wiley & Sons, Inc., 1958), p. 59.
3. Moldover, M. R., J. P. M. Trusler, T. J. Edwards, J. B. Mehl, and R. S. Davis. "Measurement of the Universal Gas Constant R Using a Spherical Acoustic Resonator," *J. Res. Natl. Bur. Stand. (U.S.)* 93:85 (1988).
4. Knudsen, M. *The Kinetic Theory of Gases,* 3rd ed., (New York: John Wiley & Sons, Inc. 1950), 3rd ed., p. 1.

5. Knudsen, M. "Maximum Rate of Evaporation of Mercury," *Ann. Phys.* 47:697 (1915).
6. Guggenheim, E. A. "Statistical Thermodynamics of Mixtures with Zero Energies of Mixing," *Proc. R. Soc.* A183:203 (1944).
7. Kittel, C. *Elementary Statistical Physics* (New York: John Wiley & Sons, Inc., 1958), pp. 169—171.
8. Hertz, H. "On the Vaporization of Liquids, Particularly Mercury, in an Evacuated Space," *Ann. Phys.* 17:177 (1882).
9. Mortensen, E. M., and H. Eyring. "Transmission Coefficients for Evaporation and Condensation," *J. Phys. Chem.* 64:846 (1960).
10. Glasstone, S., K. J. Laidler, and H. Eyring. *The Theory of Rate Processes* (New York: McGraw-Hill Book Company, 1941).
11. Glasstone, S., K. J. Laidler, and H. Eyring. *The Theory of Rate Processes* (New York: McGraw-Hill Book Company, 1941), p. 14.
12. Wyllie, G. "Evaporation and Surface Structure of Liquids," *Proc. R. Soc.* A197:383 (1949).
13. Herzberg, G. *Infrared and Raman Spectra of Polyatomic Molecules* (New York: D. Van Nostrand Company, 1945), p. 534.
14. Kincaid, J. F., and H. Eyring. "Free Volume and Free Angle Ratios of Molecules in Liquids," *J. Chem. Phys.* 6:620 (1938).
15. Hildebrand, J. H. "Liquid Structure and Entropy of Evaporation," *J. Chem. Phys.* 7:233 (1939).
16. Pitzer, K. S. "Corresponding States for Perfect Liquids," *J. Chem. Phys.* 7:583 (1939).
17. Alty, T. and C. A. Mackay. "The Accomodation Coefficient and the Evaporation Coefficient of Water," *Proc. R. Soc.* A149:104 (1935).
18. Prüger, W. "Rate of Evaporation of Liquids," *Physik* 115:202 (1949).
19. Lewis, W. K., and W. G. Whitman. "Principles of Gas Absorption," *Eng. Chem.* 16:1215 (1924).
20. Langmuir, I., and D. B. Langmuir. "The Effect of Monomolecular Films on the Evaporation of Ether Solutions," *J. Phys. Chem.* 31:1719 (1927).
21. Davies, J. K., and E. K. Rideal. *Interface Phenomena,* 2nd ed., (New York: Academic Press, Inc., 1963), p. 301.
22. Bird, R. B., W. E. Stewart, and E. N. Lightfoot. *Transport Phenomena* (New York: John Wiley & Sons, Inc., 1960).
23. Lewis, W. K., and K. C. Chang. "The Mechanism of Rectification," *Trans. AIChE* 21:127 (1928).
24. Colburn, A. P., and T. B. Drew. "The Condensation of Mixed Vapors," *Trans. AIChE* 33:197 (1937).
25. Sherwood, T. K., and R. L. Pigford. *Absorption and Extraction* (New York: McGraw-Hill Book Company, 1952), Chapter 9.
26. Brutsaert, W. "A Model of Evaporation as a Molecular Diffusion Process into a Turbulent Atmosphere," *J. Geophys. Res.* 70:5017 (1965).
27. Sutton, O. G. *Micrometeorology* (New York: McGraw-Hill Book Company, 1953).
28. Sutton, O. G. *Micrometeorology* (New York: McGraw-Hill Book Company, 1953), p. 68.

29. Kingdon, K. H. "Enhancement of the Evaporation of Water by Foreign Molecules Adsorbed on the Surface," *J. Phys. Chem.* 67:2732 (1963).
30. Carlon, H. R. "Accelerated Evaporation of Water from Moistened, Porous Substrates," In press.
31. Gittens, G. J. "Variation of Surface Tension of Water with Temperature," *J. Colloid Interface Sci.* 30:406 (1969).
32. Drost-Hansen, W. "Aqueous Interfaces. Methods of Study and Structured Properties," *Chemistry and Physics of Interfaces,* Symposium on Interfaces (Washington, DC: American Chemical Society, 1965).
33. Franks, F. *Chem. Ind.* 204 (1961).
34. Taylor, J. E. "An Apparatus for the Continuous Production of Triple Distilled Water," *J. Chem. Educ.* 37:204 (1960).
35. Butler, E. B. "Contact Angles and Interfacial Tensions in the Mercury-Water-Benzene System," *J. Phys. Chem.* 67:1419 (1963).
36. Eigen, M., and L. De Maeyer. "Investigations Concerning the Kinetics of Neutralization I," *Z. Electrochemie* 59:986 (1955).
37. MacIntyre, F. "Ion Fractionization in Drops from Breaking Bubbles," PhD Thesis, Massachusetts Institute of Technology, Cambridge, MA, 1965.
38. Smith, V. C. "Problems in the Production and Handling of Ultra Pure Water," International Water Conference, 29th Annual Meeting of the Engineers Society of Western Pennsylvania, Pittsburgh, PA, 1968.
39. Simon, G. P., and C. Calmon. "Ultrapure Water for the Semiconductor and Microcircuit Industries," *Solid State Technol.* 21, (February 1968).
40. Callaghan, T. J. "A Practical Guide for the Selection of a Water Purification System," *Am. Lab.* 60—67 (May 1968).
41. Nykanen, J. F., and R. M. Cutler. "Designing and Operating a Pharmaceutical Ultrapure Water System," *Microcontamination,* 51 (May 1990).

CHAPTER 3

Evaporation and Condensation Coefficients

INTRODUCTION

There has been controversy in the literature which can be summarized as concerning the answer to the question: "Are the evaporation coefficient and the condensation coefficient of water equal to 1, or is water anomalous in this characteristic?" In this chapter, much of the experimental work on evaporation and condensation coefficients is summarized, and the answer to this question can be found in the preponderance of work of respected workers. The misapplication of the Hertz-Knudsen equation will be treated in later sections. An early paper by Langmuir on the vapor pressure of metallic tungsten serves as background.

Langmuir[1] considered a surface of metal in equilibrium with its saturated vapor. Equilibrium is looked upon, according to the *kinetic theory*, as balance between the rate of evaporation and the rate of condensation, these two processes going on simultaneously at equal rates.

He observed that when one boils water as rapidly as one can, the *net* rate at which steam is produced is very small compared to the actual rate at which water is evaporating into steam and the steam again condensing on the water. The steam is considered to be saturated at the surface boundary between the water and the steam. At much lower temperatures at which the vapor pressure is of the order of 0.001 mm of mercury (2×10^{-5} psi), the rate of evaporation of a substance, even in a practically perfect vacuum, is very small due to the fact that the rate at which molecules are formed is limited.

At temperatures so low that the vapor pressure does not exceed 1 mm of mercury (0.02 psi), the actual rate of evaporation can be considered to be independent of the presence of vapor around it; that is, the rate of evaporation in a high vacuum is the same as the rate of evaporation in the presence of evaporated vapor. Similarly, the rate of condensation may be considered to be determined only by the pressure of the vapor.

Langmuir expressed the rate at which the vapor comes in contact with the metal as:

$$m = \sqrt{M/2\pi RT} \cdot p \qquad (1)$$

where m = the mass flux of metal
M = the molecular weight of the metal
R = the universal gas constant

T = the absolute temperature
p = the vapor pressure

If it can be assumed that every atom of the vapor which strikes the metal condenses, this equation gives the desired relation between the rate of evaporation in vacuo and the vapor pressure. If, however, a certain proportion r of the atoms of the vapor is reflected from the surface, the vapor pressure will be greater than that calculated using this equation in the ratio $1/(1 - r)$.

REFLECTION OF WATER VAPOR MOLECULES AT AN EVAPORATING SURFACE

Alty[2] endeavored to determine whether, and what percentage of, water vapor molecules are reflected at the surface of evaporating liquid. He began with an expression from the application of the kinetic theory of gases with no reflection for the mass of molecules leaving unit area of liquid surface per minute, which is also the mass of molecules striking the unit area per minute from the saturated vapor. The expression is

$$m = 14.63\ P_s/\sqrt{T_s} \tag{2}$$

where m = in g/cm²·min
P_s = the saturation vapor pressure in millimeters of mercury
T_s = the surface temperature in °K

Before this expression can be used in connection with the evaporation of water it is necessary to show that there is no reflection at the surface. The experimental work of Alty was undertaken to examine this equation as carefully as possible.

The mass of water evaporating into a vapor space at pressure p, and with a fraction f of incident vapor molecules getting into the liquid is given by:

$$m = 14.63\ (P_s - p)f/\sqrt{T_s} \tag{3}$$

The fraction f could be determined if the mass evaporating into a perfect vacuum could be measured, from the equation:

$$f = m\sqrt{T_s}/14.63\ P_s \tag{4}$$

In preliminary experimental work the surface temperature was very much reduced when the rate of evaporation was large; it was possible to freeze the water by rapid evaporation. At higher temperatures rapid evaporation produced large temperature gradients in the liquid so that it boiled instead of evaporating steadily. Subsequently, the temperature gradients were reduced as much as possible by supplying heat to the evaporating water at the greatest possible rate. This was accomplished by using a glass experimental cell with very thin walls, and immersing it in a mercury thermostat so as to increase the rate of heat transfer to the water. The cell was connected directly to a manometer and, through a leak, to a vacuum pump of high capacity. A set of these leaks, offering resistance to the flow of vapor, was so arranged that they could be interchanged with ease. The pressure, p, above the evaporating surface could be controlled in this way, and the rate of evaporation for different values of p could be measured. In order to minimize the chance of boiling, all water was freshly distilled and boiled to remove air.

The experiment was performed in two parts: (1) the temperature difference, ΔT, between the evaporating surface and the surrounding mercury was obtained as a function of p, and ΔT was plotted against p; and (2) the evaporation rate, M, was obtained as a function of p, and M was plotted against p. The plots were then extrapolated to cut the p-axis at $\Delta T = \Delta T_0$ and $M = M_0$, respectively. M_0 is the rate of evaporation into vacuum from a surface at the temperature $(T_B - \Delta T_0)$, if T_B is the temperature of the bath. Under these conditions, the preceding equation would apply if we set $T_s = (T_B - \Delta T_0)$, $m = M_0/A$, P_s = saturation vapor pressure at the temperature T_s, and A = surface area of the evaporating water. Thus, it should be possible to determine whether there is reflection at the surface.

In order to determine the surface temperature, a thermocouple was introduced into the experimental cell; the other junction was immersed in the mercury thermostat. The cell was filled up to a point and the pump was started; the temperature difference was measured every minute until the water had evaporated to a lower point. The two levels were measured by means of a reading telescope. The time at which the cell thermocouple junction was in the surface was calcuated from the total change in level, and the thermocouple reading at this time was taken as ΔT. To determine the evaporation rate, M, a weighed quantity of water was placed in the cell and evaporation proceeded for a known time so that M was determined for the loss of weight and the time.

On comparison of the experimental results with the formula for the kinetic theory of gases for the number of vapor molecules striking a water surface, it appeared that only about 1% of the vapor molecules incident on the water surface were able to enter the liquid. This was taken to indicate that there must have been very pronounced reflection of water vapor molecules at the liquid surface.

CALCULATION OF MASS FLUX

Alty[3] investigated the maximum rate of evaporation of water, initially calculating mass flux using Knudsen's formula[4] transformed to:

$$m = 43.75 \times 10^{-6}\, p\sqrt{M/T} \qquad (5)$$

where m_e = the mass flux of water vapor ($g/cm^2 \cdot sec$)
p = water vapor pressure ($dyne/cm^2$)
M = the molecular weight of water (g/mol)
T = absolute temperature

Alty stated that, if the water vapor pressure at the temperature of the experiment were sufficiently small, the mass flux would be affected only very slightly by the presence of the vapor in the surrounding space. Then the mass of saturated vapor striking unit area of surface per second would be:

$$m_s = 43.75 \times 10^{-6}\, Ee_s\sqrt{M/T_s} \qquad (6)$$

where E (the evaporation coefficient) is defined as the rate of evaporation measured experimentally to that calculated using Equation 6
e_s = the saturation vapor pressure of water
T_s = the absolute temperature of the saturated vapor

For evaporation into a space in which the vapor pressure, e, is less than saturation vapor pressure (but not zero), the net mass flux evaporated is equal to the mass flux leaving the liquid surface reduced by the mass flux returning to the liquid from the vapor:

$$m_u = 43.75 \times 10^{-6}\, E\,(e_s - e)\sqrt{M/T_s} \qquad (7)$$

The vapor pressure must be small so that the vapor present above the liquid is not sufficient to influence appreciably the rate at which the molecules leave the liquid.

The very rapid evaporation produces intense local cooling in the surface. For water at 25°C, the water surface temperature may be 20° to 30°C below that of the liquid as a whole; e_s is the saturation water vapor pressure at the temperature of the surface.

The value of E, the evaporation coefficient, was found to be 0.04 in the temperature range +4° to −8°C.

IS THE EVAPORATION COEFFICIENT UNITY?

Hickman[5] observed at the writing of his 1954 paper that little was known at that time of the actual mechanism of transfer of water molecules across the interface between liquid and vapor. His paper was intended to answer the question *whether there are conditions at which the evaporation coefficient is unity*, affirmatively, thus showing that water is not an anomalous liquid as to evaporation.

He operated on the proposition that the experimental conditions most likely to realize the maximum evaporation coefficient exist when a clean random surface of water is continuously exposed to high vacuum and instantly removed. He considered the device most likely to produce these conditions, without undue chilling or formation of ice, to be one in which a rapidly flowing stream issues from one orifice and immediately enters another. The flowing-stream tensimeter was modified in such a way that a stream of water was evaporated into high vacuum for periods ranging from a few seconds to many minutes before ice appeared.

The evaporation coefficient, *E*, defined as the ratio of the quantity of water observed evaporating to the quantity calculated, was determined from the quantity of water collected in a distant condenser. The evaporation coefficient was calculated from the formula:

$$E = w/0.0583\, p\sqrt{18\, Aft/279} \tag{8}$$

where w = the quantity of water evaporating (g/sec)
p = saturation vapor pressure of water at absolute temperature T (mm of mercury)
A = the area of evaporating surface (cm^2)
f = the escape coefficient, the ratio of the number of molecules reaching the condenser to the number of molecules actually evaporating
t = the duration of evaporation (sec)

Hickman dealt with the argument that the Langmuir formula above was originally intended to define accommodation coefficients under reversible conditions at near equilibrium and that it should not be used to predict rates because rate processes are inherently different from equilibrium processes. He maintained that an equilibrium becomes a rate process as soon as it is disturbed and that a rate process becomes an equilibrium process when it is conducted for a short enough time. He maintained that the time element in his experiments, 0.001 sec, was sufficiently small that an equilibrium equation could be applied.

The eventual back pressure in the experiments was less than 1 mm of mercury (0.02 psi). A value of escape coefficient of 0.65 ± 0.20 was used in the calculations.

Hickman concluded that a clean, new surface of water exhibits an evaporation coefficient of not less than 0.25 and that the coefficient probably approximates unity. These values were 25 to 100 times as great as those previously reported.

EXPERIMENTAL DETERMINATIONS OF EVAPORATION AND CONDENSATION COEFFICIENTS

Jamieson,[6,7] reporting on his determination of the condensation coefficient of water using a radioactive tracer technique, reviewed previous work. His experimental method involved the condensation of tritium-labeled steam on a stream of nonlabeled water which had passed through a jet at a speed of about 2000 cm/sec. The ratio of the number of steam molecules condensed, deduced from the tritium beta activity acquired by the previously nonlabeled water stream, to the number impinging on the water stream gave the condensation coefficient of 0.35 ± 0.013.

The following items are of general interest in Jamieson's review. To determine the saturation water vapor pressure, it is necessary to know the liquid surface temperature; the measurement of surface temperature remains one of the most common sources of error. Contamination of the liquid surface is another major source of error. Hickman[5] obtained a value of evaporation coefficient of not less than 0.25 by measuring the rate of evaporation of water from a high speed jet. Jamieson's work was in agreement with Hickman's result. Two possible explanations were suggested by Jamieson to explain the discrepancy between the two widely different values: (1) contamination resulted in low rates of evaporation in the experiments yielding an evaporation coefficient of less than or equal to 0.04, and (2) the value obtained depends upon the conditions under which the measurement is made. Jamieson concluded that the measured value may vary with the speed or rate of shear at the liquid surface and that nevertheless, for a static water surface, the coefficient could be about 0.04, as widely reported.

Jamieson tabulated values of evaporation coefficient for water and ice determined by other workers.

Johnstone and Smith,[8] described an investigation (which used a Mach-Zehnder interferometer[9]) of the rate of heat transfer and the temperature changes caused by rapid condensation or evaporation in several quiescent liquids and from measurements of which the values of the condensation coefficients were calculated. Johnstone and Smith mentioned that the condensation coefficient is of practical interest since it is related to the minimum value of interfacial resistance to mass transfer and it determines the maximum rate of condensation on, or evaporation from, a liquid surface. They attributed the principal difficulty in all methods of experimentally determining the condensation coefficient to the determination of the liquid surface temperature.

In their experiment, an evacuated cell was half filled with purified and deaerated liquid which was allowed to come to equilibrium under vacuum at a chosen temperature and, hence, pressure. The pressure was then altered slightly (usually increased causing condensation and therefore transfer of heat to the liquid surface) by adjusting a bellows in the cell. The resulting nonsteady change of temperature of the liquid was measured by recording the position of a diaphragm on which the liquid rested and which formed the base of the cell. Using the measured temperature change and the measured pressure change, the condensation coefficient was calculated. The value of the coefficient was found to decrease with time. The values ranged from 0.01 at long exposure times and were extrapolated to 0.2 to 0.5 at zero exposure time.

The Johnstone and Smith experimental method was distinct from other methods. The other methods (for example, the liquid jet method used by Jamieson and the dropwise condensation method which yielded high coefficient values) involved short times of exposure of liquid to vapor whereas using this method coefficients were determined at different exposure times while the liquid surface was kept quiescent. Possible explanations for the fall in the value of the calculated coefficient with time were discussed.

The authors asserted that the fact that a range of values for the coefficient is found in one experiment explains the discrepancies between the results of different investigators and that the results correspond with those obtained from liquid jets and permit accounting for the high values calculated from the results of experiments on dropwise condensation.

Hickman,[10] in a review of the evaporation coefficient, attributed the low experimental values of E to resistance to flow of heat. He noted that the temperature of the top layer of evaporating molecules determines the surface vapor pressure in the Hertz-Knudsen equation, and reviewed several attempts at measuring water surface temperature.

Referring to his own earlier work,[11] Hickman noted that increases in evaporation yield in stirred, falling-film and centrifugal stills were evidence that low values of E resulted less from chemical rearrangement than from heat "starvation", macroscopic surface contamination, or both. He concluded that by expediting heat transfer the "true natural value" of E could be made to approach the measured value. Furthermore, he proposed to expedite heat transfer by replacing the surface layers of water at a turnover rate of the order of 10^5 to 10^6/sec.

Using a flow tensimeter[12] (a falling-stream molecular still) and turnover rates of the order of 10^2 to 10^4/sec, Hickman and collaborator measured values of E as high as 0.5 for water.[5] Hickman stated that the exactness of the turnover jet method necessitates correcting for other resistances to mass transfer, notably gradients in the vapor, which he stated were first described by Fraser[13] and were fully expounded by Schrage.[14]

Eagleton and graduate students[15] undertook a reevaluation of the unsteady state method used by Delaney, Houston, and Eagleton[16] for evaporation experiments. The apparatus consisted of an enclosed volume into which a sample of liquid was enclosed. An evaporation experiment was performed by quickly lowering the pressure in the vapor space, the rate of evaporation being calculated from the rate of pressure increase in the gas volume closed to the surroundings. Uncertainty in the required true liquid surface temperature determined "plagued" this method also. Determinations of evaporation coefficient ranged from high values of 0.5 to 0.6 at short elapsed time to about 0.05 at relatively long time. Measurements were made within 0.2 sec and up to 1.4 sec after the start of evaporation. The decrease with elapsed time was very rapid. The Schrage[14] correction for the mass flux away from the water surface was used to correct the values of the coefficient, resulting in lower values. The reason for the variation of the coefficient was not established; however, it is possible that with increased period of evaporation the measured surface temperature was further from the true surface temperature as a consequence of a rapid increase in temperature gradients at the surface or of the influence of surface tension-driven convection. The condensation coefficient of ice, for which convection would not be present and for which temperature gradients near the surface would be much lower due to the higher thermal conductivity of ice, was investigated in the same apparatus. Values of the evaporation coefficient of about 0.55 were found for all values of elapsed time in the range 0.2 to 1.0 sec. No decrease with time similar to that for water was found, indicating that this decrease was not a peculiarity of the apparatus. Other types of run were also made. Reduction of the evaporating surface area by a factor of two had no effect on the coefficient. Condensation runs were made and the value of the condensation coefficient at 0.2-sec elapsed time was the same as for evaporation. Carbon tetrachloride gave high values of evaporation coefficient of 0.9 to 1.0 at low elapsed time and values below 0.1 at longer times. Eagleton et al. concluded that their evidence and that in the contemporary literature supported the view that the correct value for the evaporation coefficient for water is greater than 0.5 and might be 1.0, that apparently the coefficient for evaporation from a stagnant surface is about the same as for evaporation from a jet, and that condensation and evaporation result in the same coefficient.

Coughlin and graduate students[17] patterned apparatus after Jamieson's design (taking care to study and eliminate an undesirable boundary layer effect resulting from the specific type of jet design used by Jamieson) to investigate further the transient behavior of the condensation coefficient of water. Initial data obtained using the apparatus tended to confirm that the condensation coefficient decreases with time of exposure of the liquid water surface.

Coughlin[18] considered the experiments of Jamieson to have provided one of the few examples of transport from vapor to liquid where the transition from the penetration theory (without surface resistance) to a regime where the surface resistance becomes important and begins to control the transport rate appears evident. Coughlin examined Jamieson's data in the light of penetration theory with and without surface resistance, noting that the trend of the data suggests the penetration theory of Higbie[19] with the surface resistance to vapor-liquid transport formalized by Dankwerts[20] included. Coughlin concluded that Jamieson's experiments indicated a condensation coefficient for water at least as large as 0.4, with a possibility that the coefficient is larger, and that a hydrodynamic effect caused an apparent surface resistance to transfer corresponding to a coefficient of 0.4. The observed trend of measured coefficient with longer residence time was considered to be a coefficient which decreased with time. Coughlin referred also to the work of Johnstone and Smith[8] and Bonacci and Eagleton,[21] employing different experimental techniques, the data for which also indicated that the measured coefficient decreased with time during an experiment. Coughlin concluded that, since these latter two sets of experiments used water surfaces that had been formed some time before measurements were made, there did not yet appear to be sufficient evidence to suspect the migration of surface active impurities to the surface as the explanation of a condensation coefficient changing with time.

Maa[22] made thermal gradient calculations in an attempt to place the technique of the jet tensimeter on a quantitative basis.

He defined the evaporation coefficient as the ratio of the real exchange rate to the calculated exchange rate. The resistances to the supply of latent heat and the return of molecules from the emergent vapor limit the observed evaporation rate. The relation of the true evaporation coefficient to the apparent coefficient can be deduced if thermal gradients and vapor return can each be correctly calculated. Maa's paper is more particularly concerned with predicting the thermal gradients.

The laminar layer adjacent to the top liquid surface is the region in which any significant temperature change could occur during the time of exposure used in Maa's work. At $t = 0.001$ sec, the temperature had changed 0.4°C at 0.03 mm (0.001 in.). At 0.05 mm (0.002 in.) there was no observable change from bulk liquid temperature, confirming the concept that all the evaporation phenomena are located in a very thin sheath. The diameter of an unassociated water molecule is 3×10^{-8} cm (1.2×10^{-8} in.).

To demonstrate that a high value of apparent evaporation coefficient could be obtained by using a short exposure time and low bulk liquid temperature, various concentrations of salts were added to prevent freezing. An apparent value as 0.6 was obtained at about −15°C as theoretically predicted for pure water based on a true evaporation coefficient of one.

High apparent evaporation coefficients were measured at short exposure time and low temperature, and lower apparent coefficients were measured at long exposure time and higher temperature for both polar and nonpolar liquids. There were not significant differences in the behavior of evaporation due to differences in molecular structure or chemical properties.

Evaporation rates, using the jet tensimeter, were examined for water, isopropyl alcohol, carbon tetrachloride, and toluene at various liquid temperatures, times of exposure, and back pressures of vapor. The experimental results agreed with the thermal gradient calculations, making the assumption that the evaporation coefficient is unity, showing that (according to Maa) there was little or no resistance to molecules crossing the vapor-liquid interface in addition to the natural limitations imposed by the gas laws.

Eagleton et al.[23] extended the experimental method to the study of evaporation and condensation with water containing dissolved salts and surface contamination, noting that dissolved salts would lower the water vapor pressure and should therefore lower the rate of vaporization at a given surface temperature. Preliminary work with the apparatus indicated that the evaporation rates might be quite sensitive to water purity. The results of a computer simulation of the experiments of Bonacci and Nongbri suggested that the heat conduction mechanism alone was not sufficient to account for the amount of heat that must have reached the surface to sustain the evaporation rates observed.

The mathematical prediction of heat transfer during condensation, followed by the collection of experimental data using a condensation jet tensimeter were the subjects of a paper by Maa.[24] The collected data included leakage rates and correlation of measurement with theory.

In the jet tensimeter, a jet of well-mixed liquid with uniform temperature and clean surface was projected into its own vapor. During the exposure time of the jet surface, condensation or evaporation occurred depending on the temperature and pressure of the surrounding vapor. The data for a protected jet, rather than that for an unprotected jet, were properly compared with theory.

The agreement between data and theory was sufficiently close to justify the assumption that the condensation coefficient was approximately equal to one.

For the experiments, the exposure times of liquid jets were kept the same. The bulk liquid temperatures were kept the same in all condensation experiments, but the difference between the saturation temperature and the bulk liquid temperature and the difference between the vapor pressure and vapor pressure corresponding to the bulk liquid temperature were varied from experiment to experiment. Maa identified the overall driving forces for condensation and evaporation. The overall driving force for condensation was considered to be the summation of the driving force on the vapor side of the interface (the difference between the pressure of the vapor and the saturation vapor pressure corresponding to the liquid surface temperature) and the driving force on the liquid side of the interface (the difference between the liquid surface temperature and the bulk liquid temperature). The overall driving force for evaporation is the summation of the negative of these two quantities.

The experimentally determined condensation rates tended to be lower and evaporation rates higher than those predicted by the theoretical calculations. Possible reasons for these deviations were presented.

The condensation coefficients determined by use of the jet tensimeter agreed satisfactorily with the assumption that the condensation coefficient is equal to unity, showing again that there is little or no resistance to the molecules crossing the vapor-liquid interface in addition to the natural resistance imposed by the gas laws. As a corollary, when a vapor molecule strikes at the interface, the chance of failing to condense is small.

Maa[25] observed that the evaporation or condensation coefficient of ordinary liquids depends on: (1) the effective pressure of the vapor and (2) the true temperature of the liquid surface. The effective pressure of the vapor is complicated by the mass movement of vapor molecules to or from the liquid surface. Direct measurements of the true surface temperature could not be made because of steep thermal gradients beneath the surface.

Schrage[14] had considered the effect of mass movement of vapor molecules from or toward the vapor-liquid surface and derived a correction factor for the net rate of phase change. Maa undertook to examine and verify the Schrage correction, due to its importance in correcting calculated rates of evaporation and condensation, by simple experiments using the jet stream tensimeter. His paper also suggested a method of correction for changes in thermal skin thickness.

Maa asserted that the evaporation coefficient of ordinary liquids is unity, and that the rate of molecule exchange between a free liquid surface and its vapor at equilibrium is the same as the striking rate of vapor molecules at the interface. This is given by the kinetic theory of gases as $p_m\sqrt{M/2\pi RT_v}$, where p_m is the pressure in the vapor region, M is the molecular weight of the species, R is the universal gas constant, and T_v is the absolute temperature of the vapor region.

When the system is not in equilibrium, the net rate of evaporation or condensation, w (g/cm^2·sec), is described by the modified Hertz-Knudsen equation:

$$w = p_{tr}\sqrt{M/2\pi RT_{tr}} - \Gamma p_m\sqrt{M/2\pi RT_v} \tag{9}$$

where p_{tr} = the saturation vapor pressure corresponding to the liquid surface temperature, T_{tr}

Γ = the Schrage correction for the mass movement of the vapor

The experiments of Maa were designed to examine Schrage's method of computing the effect of mass vapor movement on the rate of phase change. The results suggested that the Schrage method of computing the rate of vapor movement to an evaporating liquid surface is satisfactory.

Determinations of the saturation vapor pressure of distilled water using a jet tensimeter showed that the jet tensimeter could be used as a convenient tool for determining vapor pressures of liquids.

The results confirmed the assumptions for the heat transfer calculation and Maa concluded that the results established with new certainty that the evaporation and condensation coefficients are unity for common liquids.

Davis et al.,[26] noting that Maa and Hickman[27] used an approximate iterative technique to estimate the surface temperature of the jet of the jet tensimeter, undertook to show that the interfacial temperature can be predicted in a rigorous manner without recourse to iterative procedures. The details of the analysis by Davis et al. are not given here; the interested reader is referred to their paper.

They applied their analysis to calculate evaporation coefficients from the data of Maa.[22] Their values of time-averaged evaporation coefficient, $\bar{E}$, were compared with the results reported by Maa. Asserting that Maa assumed the interfacial temperature to be constant, their analysis took into account the variation in the surface temperature of the jet.

For water and toluene the calculated values of $\bar{E}$ exceeded one, which effect was attributed to experimental and theoretical uncertainties in jet tensimetry. Davis et al.[26] considered that $\bar{E}$ for water and toluene is approximately unity, and they concluded that surface contamination and uncertainties in surface temperature were the most probable causes of errors in work reporting very low values of evaporation coefficient.

In response to a paper on the evaporation coefficient of water related to monolayers, Hickman[28] succinctly summarized the use by himself and colleagues of evaporation and condensation coefficients and other related matters. It will be helpful to quote from his letter to the editor.

In differing with the use of α as the apparent evaporation coefficient, Hickman said that "α purports to express the fundamental interchange between a liquid and its vapor, unique to that chemical species. The symbol a or α was adopted in the early days of vacuum technology for the coefficient of evaporation of mercury and for the adsorption or rejection of foreign molecules by clean metals, e.g., tungsten; hence, the term sticking coefficient."

To avoid confusion, Hickman and colleagues used E for evaporation coefficient and C for condensation coefficient for the complex heterogeneous systems such as the transfer of vapor from or to a lake to which a protective layer has been applied. They have attempted to prove experimentally that across a clean surface:

$$\mathrm{E} = \mathrm{C} = \frac{\text{Rate calculated, Hertz-Knudsen}}{\text{``Perfect'' physical measurement}} = 1.0 \tag{10}$$

Since Hickman considered that the interfacial temperature, T_{tr}, during displaced equilibrium could not be measured by any technique known at that time; he and colleagues had used bulk liquid temperature, T_b, with the strict limitation that T_{tr} should be as near to T_b as experimentally attainable and that the difference between the two temperatures be readily calculable. They assigned $\bar{E}$ to the best experimentally measured value of the evaporation coefficient and reserved E for the "true" value after deriving T_{tr}.

Hickman maintained that the values of E, C, or α thus determined tell something of the true interchange between a given liquid and its vapor.

HERTZ-KNUDSEN EQUATION

In the literature, there are several versions of the Hertz-Knudsen equation. The Hertz-Knudsen equation predicts the absolute upper limit on the rate at which molecules can escape from a liquid or solid interface into a perfect vacuum. In its simplest form it can be written as:[25]

$$w = p_{tr}\sqrt{(M/2\pi RT_{tr})} \qquad (11)$$

where
- w = evaporation rate in g/cm²·sec
- M = the molecular weight of the evaporating substance, 18.0152 g/mol for water
- R = the universal gas constant, 8.314471×10^7 joules/K·mol
- T_{tr} = the absolute temperature in kelvins, K, at the interface
- p_{tr} = the saturation vapor pressure, in dyne/cm², at T_{tr}

Equation 11 can also be written as:

$$w = p_{tr}\sqrt{(1/2\pi mk_B T_{tr})} \qquad (12)$$

where
- m = the mass of a molecule of the evaporating substance, 2.99150×10^{-23} g for water
- k_B = the Boltzmann constant, 1.38047×10^{-16} erg/deg

The net rate of evaporation in a partial vacuum has been expressed[21] as:

$$w = E\sqrt{(M/2\pi R)}\,(p_{tr}/\sqrt{T_{tr}} - p_{min}/\sqrt{T_v}) \qquad (13)$$

where
- E = the evaporation coefficient, dimensionless
- p_{min} = the vapor pressure in the vapor region
- T_v = the absolute temperature in the vapor region

Example — In this example, the calculation of evaporation rate and the determination of the evaporation coefficient will be illustrated. The maximum evaporation rate is calculated using Equation 11; Equation 12 would give the same result.

$$w = p_{tr}\sqrt{(M/2\pi RT_{tr})}$$

At 20°C, $T_{tr} = 273.15 + 20 = 293.15$ K; $p_{tr} = 17.535$ mm of mercury = 23,378 dyne/cm^2; M = 18.0152 g/mol, and R = 8.314471 × 10^7 joules/K mol.

$$w = 23{,}378\sqrt{18.0152/2\pi \times 8.314471 \times 10^7 \times 293.15}$$
$$w = 23{,}378\sqrt{1.11763 \times 10^{-10}} = \mathit{0.254}\ \text{g/cm}^2\cdot\text{sec}$$

Consider now that the measured value of w, determined by measuring the loss of weight (mass) of a body of water of known surface area for a specified length of time, is 0.010 g/cm^2·sec. Using now the definition of evaporation coefficient, *E:*

E = measured experimental evaporation rate/calculated maximum evaporation rate

E = 0.010/0.254 = *0.039*

MAXIMUM EVAPORATION RATE FOR WATER

There has been controversy in the literature concerning the use of the Hertz-Knudsen equation to predict evaporation rates of liquids, particularly water, at atmospheric pressure. The controversy will not be joined here. Rather, the conditions under which the Hertz-Knudsen equation can be used will be examined using several pertinent references.

de Boer in his book, *The Dynamical Character of Adsorption,*[29] defines the maximum rate of evaporation very clearly: "If we now maintain the temperature of a liquid at a constant value and we *take all the vapour away and keep taking it away* (italics added), preventing any molecule from returning to the liquid, we can immediately calculate how much liquid will evaporate per cm^2 and per second. This is obviously the maximum rate of evaporation which we can ever attain at the chosen temperature."

The maximum rate of evaporation of water (g/cm^2·sec) can be expressed as:

$$v_{max} = 0.0583\ p\sqrt{(18/T)} \qquad (14)$$

where p = saturation vapor pressure (expressed in mm of mercury) of water at absolute temperature, T

This equation holds only if the water molecules which strike the surface are not reflected at the surface; if some of the molecules are reflected at the surface, the rate of evaporation will be smaller.

Under these conditions, at 20°C, water would evaporate at the rate of 0.253 g/cm^2·sec which corresponds to 9 m/hr (30 ft/hr). In reality, the evaporation rate of water is smaller by a factor of 100,000 to 1,000,000. A tropical sea evaporates at a rate of about 2 m/year (6.6 ft/year), in more moderate climates the rate is about 50 cm/year (1.6 ft/year). If water evaporated at the maximum rate, many lakes and seas would evaporate completely in a few hours, and the water of the oceans would evaporate completely in a number of days. For example, Lake Mead in Nevada would evaporate completely in about 6 hr.

de Boer considered the slow actual rate of evaporation to be due to the fact that there is a thin gas layer over the liquid surface in which the water vapor pressure is near saturation. The slow diffusion of water vapor from this layer into the air above was thought to mainly govern the actual rate of evaporation. This is in addition to the fact that the temperature of the water would decrease during evaporation.

Sherwood, Pigford, and Wilke[30] gave a particularly clear discussion of the limiting transport rate of a gas. The discussion here will follow their description closely.

Consider a system in which water and pure water vapor are in contact and in equilibrium. The gas pressure is equal to the vapor pressure of water at the temperature of the liquid surface, and there is no net transport from one phase to the other. The gas molecules move at high speed, collide with the liquid surface, and are incorporated. To maintain equilibrium, evaporation must occur at the same rate. The rate of collision of the vapor molecules with the surface is calculated from kinetic theory; some fraction α of these remains in the liquid, the fraction $(1 - \alpha)$ rebounds to the gas. One may conclude that the rate of evaporation into an absolute vacuum must be α times the rate of collision from the gas saturated at the temperature of the surface, assuming that the rate of evaporation is not influenced by the pressure of gas.

Hertz,[31] Knudsen,[4] and Langmuir[1] were led by this reasoning to the following expression for the maximum possible rate of transport from a surface to the gas:

$$\begin{aligned} N_A &= 1006\ \alpha\ (2\pi M R T_s)^{-1/2}(p_s - p_G) \\ &= 44.3\ \alpha\ (M T_s)^{-1/2}(p_s - p_G) \end{aligned} \qquad (15)$$

where N_A is in g mol/cm^2·sec

M = the molecular weight of the water

R = the universal gas constant

T_s = the water surface temperature

p_s = the saturation water vapor pressure at the water surface temperature

p_G = the vapor pressure in the gas phase

The temperature is in °K and the pressure is in atmospheres.

This maximum transfer rate is very large in comparison with mass-transfer rates encountered in most industrial equipment. If α were equal to unity, a free-water surface at 20°C evaporating into an absolute vacuum would retreat at the rate of 2.6 mm/sec. Only in high-vacuum and space technology is such a rate encountered.

Some of the overall potential is used in achieving the phase change. Thus, the gas immediately in contact with the water surface is not in equilibrium with it. The resistance to transport across the gas-water interface, $1/k_i$ (where k_i is the mass-transfer coefficient for transport across the interface), is given by:

$$1/k_i = \sqrt{(2\pi MRT_s)}/(1006\ \alpha) = 0.0225\sqrt{(MT_s)}/\alpha \qquad (16)$$

The first part of Equation 15 could be written:

$$N_A = k_i\,(p_s - p_G)$$

The units of k_i are g mol/cm²·sec·atm. For a water surface at 20°C, with $\alpha = 1$, the magnitude of k_i is 0.612. Such a small resistance may amount to several percent of the total resistance to transport if the transport rate is large, as in vacuum. The interfacial barrier is generally quite negligible if the mass-transfer rate is small.

The limiting rate given by the Hertz-Knudsen equation becomes important in practice only when the transfer rates are exceptionally high.

The coefficient α is called the evaporation coefficient, the sticking coefficient, or the accommodation coefficient. There is no useful theory for predicting α, and it cannot be easily experimentally determined. The experimental determination requires measurement of the surface temperature; this leads to significant errors since the temperature gradient near the surface can be very steep. Littlewood and Rideal[32] and others questioned the validity of most of the published values of α because of questionable surface temperature measurements. Maa,[33] using a method not requiring a probe to measure the surface temperature, obtained values of α of approximately 1.0 for water.

Sherwood et al.[30] considered it to be conceivable that most of the published values of α were in error, and that α is essentially 1.0 for all simple liquids including water.

Palmer,[34] in response to another published paper,[35] reviewed the subject of retardation of evaporation by monolayers. The following are quotations of Palmer's remarks.

A clear understanding is needed of how monolayers reduce evaporation rates and of how to predict the magnitude of these reductions from fundamental physical principles.

The Hertz-Knudsen equation . . . predicts the absolute upper limit on the rate at which molecules can escape from a liquid (or solid) interface into a perfect vacuum.

The evaporation coefficient then is used to account for deviations from kinetic theory that may be due to anomalous molecular interactions at the interface which are exposed in the dynamic state and which are not properly accounted for in thermodynamic properties.

It is currently suspected that no such deviations exist and that the evaporation coefficient is essentially unity for all simple liquids.

Because this maximum evaporation rate is so large relative to mass transfer rates normally encountered, such as in the measurement of evaporation resistances due to insoluble monolayers, effects arising from kinetic theory arguments are negligible and are not considered when the usual resistances to diffusion and heat transfer are present in a system.

It is only when experiments are devised or unusual circumstances are encountered in which these high mass transfer rates are approached, that the influence of the kinetic energy of molecules and the evaporation coefficient can even be detected (such as evaporation into high vacuum or space technology).

Consider the evaporation of water with a surface temperature of 300°K, assuming that the true evaporation coefficient is unity. From the Hertz-Knudsen equation the evaporative flux into a perfect vacuum . . . is 0.38 $g/cm^2 \cdot s$. On the other hand, an estimate for the evaporative flux of water into dry air at atmospheric pressure may be obtained from a film theory diffusion model assuming . . . that the only resistance to evaporation is due to diffusion in the gas phase (heat transfer effects neglected). With a 1-mm (0.04-in.) thin boundary layer *d* and a molecular diffusivity equal to 0.26 cm^2/s.

The evaporative flux is 6.7×10^{-5} $g/cm^2 \cdot sec$.

The two results are dramatically different because a substantial diffusional resistance in the gas phase exists in the latter case — *not* because of any anomaly associated with molecular passage through the interface.

Plesset[36] observed that although the evaporation coefficient and the condensation coefficient have the same value under equilibrium conditions, the assumption that they have the same value under nonequilibrium conditions may not be justified.

REFERENCES

1. Langmuir, I. "The Vapor Pressure of Metallic Tungsten," *Phys. Rev.* 2:329—342 (1913).
2. Alty, T. "The Reflection of Molecules at a Liquid Surface," *R. Soc. Proc.* 131:555—564 (1931).
3. Alty, T. "The Maximum Rate of Evaporation," *Philos. Mag.* Ser. 7. 5:82—103 (1933).
4. Knudsen, M. *Ann. Phys.* 47:697 (1915).
5. Hickman, K. C. D. "Maximum Evaporation Coefficient of Water," *Ind. Eng. Chem.* 46:1442—1446 (1954).
6. Jamieson, D. T. "Condensation Coefficient of Water," *Nature* 202:583 (1964).
7. Jamieson, D. T. "The Condensation Coefficient of Water," in *Proceedings of the ASME Symposium, Advances in Thermophysical Properties at Extreme Temperatures and Pressures*, Purdue University, Indiana (1965).
8. Johnstone, R. K. M., and W. Smith. "Rate of Condensation or Evaporation During Short Exposures of a Quiescent Liquid," in *Proceedings of the Third International Heat Transfer Conference* (New York: American Institute Chemical Engineers, 1966) 2:348.
9. Johnstone, R. K. M., and W. Smith. "A Design for a 6-in. Field Mach-Zehnder Interferometer," *J. Sci. Instrum.* 42:231 (1965).
10. Hickman, K. "Reviewing the Evaporation Coefficient," *Desalination* 1:13—29 (1966).
11. Hickman, K. *Chem. Rev.* 34:51 (1944).
12. Hickman, K., and D. Trevoy. *Ind. Eng. Chem.* 44:1882 (1952).
13. Fraser, R. G. J. *Molecular Rays* (London: Cambridge Univ. Press, 1941).
14. Schrage, R. W. *A Theoretical Study of Interphase Mass Transfer* (New York: Columbia University Press, 1953) p. 92.
15. Eagleton, L. C. *Saline Water Conversion Report for 1967*. U.S. Department of Interior, p. 91.
16. Delaney, L. J., R. W. Houston, and L. C. Eagleton. "Rate of Evaporation of Water and Ice," *Chem. Eng. Soc.* 19:105 (1964).
17. Coughlin, R. W. *Saline Water Conversion Report for 1967*, U.S. Department of Interior, p. 89.
18. Coughlin, R. W. "Surface Resistance in Transport from Vapor to Liquid," *Ind. Eng. Chem.* 22:1503 (1967).
19. Higbie, R. "The Rate of Absorption of a Pure Gas Into a Still Liquid During Short Periods of Exposure," *Trans. Am. Inst. Chem. Eng.* 31:365 (1935).
20. Dankwerts, P. V. "Significance of Liquid Film Coefficients in Gas Absorption," *Ind. Eng. Chem.* 43:1460 (1951).
21. Bonacci, J. C., and L. C. Eagleton. "Maximum Rate of Evaporation and Condensation of Water," presented at the U.S. Office of Saline Water Symposium, Rochester, NY (November 1966).
22. Maa, J. R. "Evaporation Coefficient of Liquids," *Ind. Eng. Chem.* 6:504—516 (1967).
23. Eagleton, L. C. *Saline Water Conversion Report for 1968*, U.S. Department of Interior, (1968), p. 275.

24. Maa, J. R. "Condensation Studies with the Jet Stream Tensimeter," *Ind. Eng. Chem. Fundam.* 8:564—570 (1969).
25. Maa, J. R. "Rates of Evaporation and Condensation between Pure Liquids and Their Own Vapors," *Ind. Eng. Chem. Fundam.* 9:283—287 (1970).
26. Davis, E. J., R. Chang, and B. D. Pethica. "Interfacial Temperatures and Evaporation Coefficients with Jet Tensimetry," *Ind. Eng. Chem.* 14:27—33 (1975).
27. Maa, J. R., and K. Hickman. *Desalination* 10:95 (1972).
28. Hickman, K. "Comments on a Paper by Barnes," *J. Colloid Interface Sci.* 65:573 (1978).
29. de Boer, J. H. *The Dynamical Character of Adsorption* (London: Clarendon at the University Press, 1953), Chapter 2.
30. Sherwood, T. K., R. L. Pigford, and C. R. Wilke. *Mass Transfer* (New York: McGraw-Hill Book Company, 1975) pp. 178—184.
31. Hertz, H. *Ann. Phys. Chem.* 17:177—198 (1882).
32. Littlewood, R., and E. Rideal, *Trans. Faraday Soc.* 52:1598 (1956).
33. Maa, J. R. *Ind. Eng. Chem. Fundam.* 6:505 (1967); 9:283 (1970).
34. Palmer, H. J. "Re: 'Insoluble Monolayers and the Evaporation Coefficient of Water' by G. T. Barnes," *J. Colloid Interface Sci.* 65:574—575 (1978).
35. Barnes, G. T. "Insoluble Monolayers and the Evaporation Coefficient of Water," *J. Colloid Interface Sci.* 65:566—572 (1978).
36. Plesset, M. S. "Note on the Flow of Vapor Between Liquid Surfaces," *J. Chem. Phys.* 20:790 (1952).

CHAPTER 4

Evaporation from Water Drops

INTRODUCTION

Evaporation of water from water drops has been of interest for some time to some hydrologically oriented investigators and to atmospheric scientists (in investigations of precipitation mechanisms, for example).

KINETICS OF EVAPORATION OF DROPLETS

Fuchs (Fuks)[1] undertook a complete survey devoted to the kinetics of evaporation and growth of droplets of pure liquids, a survey of the at-time published experimental work on droplet evaporation with selected theoretical work which could be compared with experimental work. The survey included 88 references. The following statement from his preface outlines the evaporation process for droplets: "Under natural conditions this phenomenon is extremely complex. The bulk of the droplet evaporates almost immediately; the process is non-stationary and occurs in a medium with unequal temperature and vapour concentration; the drops move irregularly relative to the medium and are more or less deformed, while circulation arises within the drops; heat transfer between the drops and the medium occurs by three different mechanisms (conduction, convection and radiation)."

EVAPORATION FROM FREELY FALLING WATER DROPS

Kinzer and Gunn[2] considered, theoretically and experimentally, evaporation from freely falling water drops moving at terminal velocity relative to the environmental air. The theoretical approach involved the calculation of equilibrium evaporation rates by two independent processes, one depending upon the laws of heat transfer and the other concerned with the transfer (under the influence of vapor density gradients) of water vapor outward from the drops. The equations used to calculate the rate of evaporation for drops falling freely through a known environment are

$$-(dm_o/dt) = 4\pi a\rho D(X_a - X_b)\,[1 + F\,(Re\,\mu/4\pi\rho D)^{1/2}] \tag{1}$$

and

$$(X_a - X_b) \simeq (\rho_a - \rho_b)/\rho - \bar{\rho}_v(T_b - T_a)/\rho\bar{T} \tag{2}$$

where dm_o/dt = the time rate of change of the mass of the drop
a = the radius of the drop (spherical)
ρ = the density of ambient air
D = the molecular diffusion coefficient for water vapor in air
X = the ratio of the vapor density to the environment density the subscripts a and b refer to the surface of the drop and to the environment into which the drop is suddenly introduced, respectively
F = a "ventilation factor", which measures the ratio of the evaporation rate for a ventilated drop to the rate for a drop at rest relative to its environment
Re = the Reynolds number (defined by $Re \equiv 2\rho aV/\mu$), where V is the ventilation velocity and μ is the viscosity of the ambient air
$\overline{\rho_v}$ and $\overline{T}$ = the average vapor density and the average absolute temperature in the transition layer of vapor and temperature, respectively

The equation of heat transfer to a moving drop is

$$-Q_v(dm_o/dt) = -dQ/dt = 4\pi ac_p\rho k(T_b - T_a) \quad (3)$$

where Q_v = the latent heat of vaporization
dQ/dt = the rate at which heat is transferred outward from the drop
c_p = the specific heat at constant pressure of the environment
k = the thermal diffusivity of the environment

Equating the heat transferred toward the drop to that carried away by the latent heat, Kinzer and Gunn arrived at the psychrometric equation for a freely falling, ventilated drop:

$$(\rho_a - \rho_b) = [(\rho c_p k/Q_v D)(1 + f/k^{1/2})/(1 + f/D^{1/2}) + \overline{\rho}_v/\overline{T}](T_b - T_a) \quad (4)$$

where f = $Fa/(\pi t_v)^{1/2}$
t_v = the transit time for a particle moving across the spherical drop (at the velocity of ventilation) defined by $t_v \equiv 2a/V$

The primary purpose of the experimental investigation of Kinzer and Gunn was the measurement of the evaporation of water drops freely falling relative to the surrounding air. Another of the objectives was the measurement of the temperature of the freely falling single water drops. The evaporation of the water drops was first determined by two methods in which the terminal velocity with time gave, through the dependence of the mass of a freely falling drop upon its terminal velocity, the rate of loss of mass. In one method, a series of flash images on a photographic negative provided a record of the evaporation of falling drops of diameter ranging from 0.001 to 0.014 cm (0.0004 to 0.0055 in.). The second method used electronic instrumentation to measure terminal velocities of charged water drops of equivalent diameter ranging from 0.004 to 0.1 cm (0.002 to 0.04 in.). Two other methods were used for making measurements on larger droplets. In the first of these, drops of equivalent diameter in the range 0.08 to 0.45 cm (0.03 to 0.18 in.) were freely supported on an airstream in a tapered tube. In the second method, drops of equivalent diameter in the range 0.40 to about 0.55 cm (0.16 to 0.22 in.) were supported on a rising column of air.

The validity (demonstrated by temperature measurements described later) of the combination of the assumption that the temperature of the freely falling water drop is the same as that of a well-ventilated wet bulb in the same environment ($\pm 0.3°C$) with the assumption that the water vapor density at the surface of the drop is the saturation water vapor density at the temperature of the drop permitted a convenient representation of the evaporation data. Setting the quantity in the brackets in Equation 1 equal to $(1 + Fa/s')$ where s' is equal to $(\pi D t_v)^{1/2}$, and noting that $(T_b - T_a)/\bar{T}$ was of the order of 1%, Equation 1 was approximated by:

$$(dm_\sigma/dt)/4\pi aD(\rho_a - \rho_b) = 1 + Fa/s' \tag{5}$$

Kinzer and Gunn plotted the measured values for the left side of Equation 5 against a. Of more than passing interest is the observation that, in the overlapping range of the photographic method and the electrical method, such a plot showed no effect of drop charging on evaporation for drop diameters down to 0.007 cm (0.003 in.).

The equilibrium temperature of freely falling water drops was measured by allowing the drops to fall into a glass cell containing temperature-controlled water. A schlieren optical system was used to detect variations in index of refraction due to temperature differences between the drop and the water. The absence of such variations was taken to indicate that the drop temperature was the same as that to which the water in the cell was adjusted. It was claimed that the method was capable of detecting temperature differences to within 0.3°C and that through this method it was "demonstrated that the equilibrium temperature is identical with the corresponding temperature of the ventilated wet bulb to within $\pm 0.3°C$."

EFFECTS OF INSOLUBLE FILMS

Snead and Zung[3] made experimental studies of the effects of insoluble films on the rates of evaporation of charged droplets of water and ethylene glycol. Rates of evaporation were followed by observing changes in the voltage required to suspend the droplet between the plates of a Millikan oil drop type apparatus as a function of time. The range of droplet radius for water-n-decanol dispersions was about 1 to 5 μm (3.9×10^{-5} to 2.0×10^{-4} in.). The relative humidity next to the Millikan apparatus was 40 to 60% RH. Values of evaporation rate for the water-n-decanol dispersion droplets were lower by about three orders of magnitude than literature values for pure water droplets somewhat larger in radius. The evaporation rates were given as the time, t, rate of change of droplet radius squared, a^2, determined from plots of a^2 vs t. Pure water droplets evaporated too rapidly to permit quantitative measurements to be made. Snead and Zung expressed the likelihood that Gudris and Kulikowa,[4] in measuring much lower evaporation rates for small droplets, were studying droplets evaporating under conditions of near saturation or saturation and that the initial fast stage of the evaporation of droplets had already occurred. Snead and Zung concluded that the experimental evidence indicated that the evaporation rates were always reduced by insoluble films on the droplets.

CHARGED WATER DROPS

Berg and George[5] described investigations of charged water drops suspended in a nonuniform ac field. They attributed the invention of the nonuniform ac field means of suspension to Straubel.[6] The apparatus consisted of a chamber in which two spheres connected together and to one end of a transformer winding with the other end connected to ground were located at the axis of a grounding cylinder in which holes permitted introduction of the drop and illumination and observation. The drop of initial diameter of about 100 μm (0.0039 in.), charged in the electric field between a syringe needle held at +3500 V and a ring-shaped grounded electrode, was introduced through a hole at 45° to the optical axis of a microscope, motion picture camera, and projector. The drop was elevated to the desired position by the application of a dc voltage. The average evaporation rates in terms of the time rate of change of drop radius squared under relative dry conditions were 2.40×10^{-6} cm²/sec at 23°C, and 0.86×10^{-6} cm²/sec at 0°C. The activation energy for evaporation was calculated to be 7.6 kcal/mol. The authors concluded that a water drop may evaporate in spurts as reported by Kingdon[7] and others, that the evidence indicates that ions are ejected from an unstable drop, that a burst of ions occurs in the evaporation of a stable drop under the experimental conditions, and that the removal of water molecules from the drop is not part of the rate-determining step.

Berg et al.[8] used the nonuniform ac field suspension method to investigate the temperature of strongly charged water droplets and concluded that the evidence showed that warm, strongly charged 100-μm diameter (0.0039-in.) water droplets retained essentially the initial temperature throughout the duration of the experiment.

VERTICAL TUNNEL STUDIES

Hoffer and Mallen[9] described a vertical tunnel for studies of small droplets in the diameter range 50 to 200 μm (0.0020 to 0.0079 in.). The tunnel consisted of three sections: a diffuser, a profile-forming section, and a transparent observational section. Bottled breathing air compressed over water, with a consistently zero Aitken count, was used as the source of clean air. The profile of the vertical component of the air velocity in the observational section was determined by the use of a heated thermistor probe. Turbulence in the tunnel was below the detection capability of the measuring techniques. At an air relative humidity of 50%, droplets were supported for 4 min by reducing the flow rate as the droplet diameter decreased by evaporation from 150 to 50 μm (0.0059 to 0.0020 in.). During evaporation, a water droplet could be positioned to within ±2 mm (±0.08 in.) in the vertical and remained to within ±2 mm (±0.08 in.) in the horizontal.

EVAPORATION LOSSES FROM SPRINKLER IRRIGATION SYSTEMS

Showalter[10] developed a simple relationship for calculating the capacity of air to absorb water droplets evaporated into it. A water droplet quickly assumes a surface temperature equal to the wet-bulb temperature of the air. The evaporative capacity of air, E_a (g/m^3), was defined by the equation:

$$E_a = (p \cdot \Delta t_w)/(7T_w) \tag{6}$$

where p = the atmospheric pressure in mb (1013.25 mb = 14.69595 psi)
Δt_w = the wet-bulb depression (air temperature − wet-bulb temperature) in °C
T_w = the wet-bulb temperature in °K (°K = °C + 273.15)

The data are relevant for determining water losses during sprinkler irrigation, the discrepancy between radar observed liquid water contents and ground-based rainfall, and other factors important in the hydrologic balance of the atmosphere. The impact of Showalter's discussion "will, hopefully, be a sudden awakening to the importance of the parameter, wet-bulb depression, which is directly useful to and easily measured by either scientist or layman."

Experiments were conducted by Yazar[11] to determine the relationships between evaporation losses from sprinkler irrigation systems and the factors affecting them. Losses from sprinkler irrigation, which may amount to a considerable portion of the amount of water discharged from sprinklers in arid and semiarid areas, depend on the climatic and operating conditions. A portable direct indicating bridge was used to measure the electrical conductivities of both the water supplied to the sprinklers and that collected in catch funnels.

Evaporation losses were determined from the relationship:

$$E = 100\,(EC_s - EC_l) \tag{7}$$

where E is the evaporation loss (%)
EC_s and EC_l are the electrical conductivities of the samples of water in the catch funnels and of the supply water, respectively

Fifty-nine evaporation-loss tests were conducted. During the tests, the wind speed ranged from 0.91 to 6.71 m/sec, water vapor pressure deficit ranged from 2.93 to 33.14 millibars, air temperature ranged from 18.9 to 36.7°C, and operating pressure on the sprinklers ranged from 207 to 317 kPa. Evaporation loss ranged from 1.5% at a vapor pressure deficit of 2.93 millibars and wind velocity of 1.34 m/sec, to 16.8% at a vapor pressure deficit of 31.16 millibars and a wind velocity of 6.26 m/sec.

A multiple regression analysis of the data indicated that wind velocity and vapor pressure deficit were the predominate factors affecting evaporation from the sprinkler sprays.

Kincaid and Longley[12] developed a model for predicting evaporation and temperature changes in water drops traveling through air and evaluated it with laboratory data. The evaporation-temperature model was comprised of three equations: the first expressed the change in drop diameters over a time increment; the second expressed the sensible heat transfer; and the third expressed the change in drop temperature over the time increment.

Experimental measurements of evaporation were made in a small wind tunnel which produced air velocity of up to 10 m/sec, comparable to velocities of droplets leaving spray-type sprinkler heads. The room housing the wind tunnel had controlled air temperature and humidity. Steady-state temperature was measured using thermocouples and dry-bulb thermometers; humidity was determined using an aspirated electric psychrometer. Air velocity was measured immediately upstream of the water droplet with a thermal anemometer sensor. The internal temperature of the droplets was measured with a copper-constantan thermocouple, 0.05 mm (0.002 in.).

The measurement of the evaporation rate for a single drop involved suspending a drop in an airstream and noting the change in droplet diameter over time as evaporation took place. A technique for measuring drop volume was used to determine drop diameter. Initial droplet diameters ranged from 0.3 to 2.5 mm (0.01 to 0.098 in.).

It was not possible to measure temperature change and evaporation simultaneously because drop temperature changed within a few seconds. A model combining heat transfer and mass diffusion equations predicted that droplets approach and actually drop slightly below the wet-bulb temperature. Temperature measurements agreed with the model quite closely. Measured loss rates were slightly higher than rates computed using the model.

Experimental results showed that the relationship of the water supply temperature to the wet-bulb temperature of the air is important in determining evaporation from sprinkler droplets. A feature of the model accounted for droplet temperature changes throughout the flight period, significantly increasing the accuracy of evaporation predictions.

REFERENCES

1. Fuchs (Fuks), N. A. *Evaporation and Droplet Growth in Gaseous Media* (New York: Pergamon Press, Inc., 1959), R. S. Bradley, Ed., translated from Russian by J. M. Pratt.
2. Kinzer, G. D., and R. Gunn. "The Evaporation, Temperature and Thermal Relaxation-Time of Freely Falling Water Drops in Stagnant Air," *J. Meteorol.* 6:243 (1949).
3. Snead, C. C., and J. T. Zung. "The Effects of Insoluble Films upon the Evaporation Kinetics of Liquid Droplets," *J. Colloid Interface Sci.* 27:25 (1968).
4. Gudris, N., and L. Kulikowa. "The Evaporation of Small Water Drops," *Z. Phys.* 25:121 (1924).
5. Berg, T. G. Owe, and D. C. George. "Investigations of Charged Water Drops," *Mon. Weather Rev.* 95:884 (1967).
6. Straubel, H. "The Stabilization of Electrically Charged Particles in Alternating Fields," *Acta Phys. Austriaca, Vienna* 13:265 (1960).
7. Kingdon, K. H. "Enhancement of Evaporation of Water by Foreign Molecules Adsorbed on the Surface," *J. Phys. Chem.* 67:2732 (1963).
8. Berg, T. G. Owe, T. A. Gaukler, and R. J. Trainor. "The Temperature of Strongly Charged Water Droplets," *J. Atmos. Sci.* 26:558 (1969).
9. Hoffer, T. E., and S. C. Mallen. "A Vertical Wind Tunnel for Small Droplet Studies," *J. Appl. Meteorol.* 7:290 (1968).
10. Showalter, A. K. "Evaporative Capacity of Unsaturated Air," *Water Resour. Res.* 7:688—691 (1971).
11. Yazar, A. "Evaporation and Drift Losses from Sprinkler Irrigation Systems Under Various Operating Conditions," *Agric. Water Manage.* 8:439—449 (1984).
12. Kincaid, C. D., and T. S. Longley. "A Water Droplet Evaporation and Temperature Model," *Trans. ASAE* 32:457—463 (1989).

CHAPTER 5

Surface Tension, Convection, and Interfacial Waves

INTRODUCTION

In this chapter, surface tension at the water-air or water-water vapor interface, convection in the liquid water, waves on the water surface, and related subjects are reviewed.

SURFACE TENSION

At the surface of an evaporating body of water, measurements of surface tension are in most cases measurements of interfacial tension. The interface of interest is the air-water or the water vapor-water interface. The usual description of the surface tension or interfacial tension as being due to the net attraction of molecules in the surface or interfacial region into the bulk liquid with a tendency to minimize the surface area is sufficient for the purposes of this discussion. The surface tension, γ, is the work done on the system to produce a unit increase in interfacial area, σ. Therefore:

$$\gamma = \partial G_s/\partial \sigma \qquad (1)$$

where G_s is the interfacial energy or surface energy

Drost-Hansen,[1] in a two-part review of aqueous interfaces, discussed the surface tension of water and reviewed work in this area, including his own. In his review he made observations of particular pertinence to the present discussion. He observed that water has a very high surface energy, consequently is readily contaminated, usually resulting in a lowering of surface tension; and that the more elaborate the equipment for very accurately determining the surface tension for water the more likely the introduction of impurities, resulting in errors in the so-called "standard" values for water. He reviewed the studies of the temperature variation of the surface tension of water and discussed at length the anomalies or "kinks" in data reported by many investigators. This subject will be discussed later. He concluded that "it is safe to insist that the observed anomalous temperature and concentration dependencies of the surface and interfacial tension of water and aqueous solutions are real; likewise that the surface tension of pure water is apparently a very complicated function of temperature."

The problem of dynamic surface tension was discussed for both the experimental and the theoretical points of view. For the case of aqueous solutions, "unsteady state" surface tension values were attributed to the lag due to the finite rate of diffusion of surface-active agents, in consequence of which a freshly formed surface would generally not have the same surface tension as that of the surface after sufficient time had elapsed for equilibrium to have been established. The time scale of interest is apparently of the order of 1 msec. Three methods of measuring the surface tension of a liquid on this time scale were discussed: (1) the oscillating jet method based on the hydrodynamic analysis of a stream from an elliptic orifice (described by Addison[2,3]) and the calculation of surface tension, for surfaces from 2 to 20 msec old; (2) the study of the bell-shaped sheet of liquid formed by the vertical head-on collision of two streams of liquid; and (3) the surface potential method in which the interfacial electric potential between the air and the liquid surface is monitored.

Pallasch,[4] using the bell method, found that at speeds from the exit tube of the apparatus of less than 200 cm/sec values of surface tension equivalent to the static values were obtained; at speeds of about 300 cm/sec the surface tension values seemed to increase linearly with speed. Jamieson[5] had referred to Göring's[6] measurements of surface tension by a dynamic method in which a vertical jet of water impinging on a horizontal circular plate formed a film nearly spherical in shape, and noted that the dynamic surface tension values deduced from the shape of the bell varied with position on the bell with values of 100 to 150 dyne/cm at the bell, decaying to the static value of 72 dyne/cm with a relaxation time of 0.001 sec.

Drost-Hansen tabulated values of dynamic tension, observed by various investigators, of 81 and 180 dyne/cm at 12°C and 83 and 87 dyne/cm at 20°C. Tabulated calculated values ranged from 109 to 206 dyne/cm at 20°C. He discussed the problem of structural orientation near a liquid interface and mentioned that estimates of the number of oriented layers near the surface of a polar liquid ranged from 1 to 1000.

Claussen[7] reanalyzed earlier data on water surface tension[8] and some more recent data obtained in his laboratory on the basis of Eötvös[9] expanded by Ramsay and Shields.[10] Of critical importance in the analysis is the concept of an amount of surface, s, over which always the same number of molecules of liquid is distributed. The value of s was taken to be equal to $(Mv)^{2/3}$, where M is the molecular weight of the liquid and v is the volume of 1 g of the liquid, i.e., the specific volume. Claussen formulated the equation:

$$\gamma\rho^{-2/3} = \Delta G_s = \Delta H_s - T\Delta S_s \qquad (2)$$

where γ = the surface tension (dyne/cm)
ρ = the bulk liquid density (g/cm^3)

ΔG_s = the surface free energy (erg) of the number of molecules taken to be the number distributed on 1 cm^2 at 4°C on the surface, s (taken to be 1 cm^2)
ΔG_s is also the free energy of surface formation
ΔH_s = the heat of surface formation
ΔS_s = the entropy of surface formation

Claussen plotted $\gamma\rho^{-2/3}$ against T (°K) over the temperature range 0 to 50°C for the data of Teitel'baum et al.,[11] Moser,[8] and Claussen.[7] He also plotted $\gamma\rho^{-2/3}/T$ against 1/T for the same data. In both cases the plots were linear. The smooth linear plot of $\gamma\rho^{-2/3}$ against T was in contrast with the "kinks" in the surface tension vs temperature plots of Drost-Hansen and others.

Claussen also commented on the attributing of 19.5 square angstrom surface area per surface molecule (a figure characteristic of many investigations of fatty acid films, being approximately the area of one fatty acid molecule adsorbed on the surface) by Grastella[12] to a characteristic of water structure. Claussen stated that in his opinion this fact had been neglected, that not only the area of the fatty acid molecule but also the water structure contributes to this common area determination, that the use of film pressure data may permit an accurate assessment of the dimensions of surface water structure, and that by this means one may be able to determine the proportionality factor between thermodynamic properties for one 4°C-cm^2 of surface and 1 mol of surface, thus permitting the calculation of precise molar surface properties.

Gittens[13] measured the surface tension at small temperature intervals, of about 0.5°C contrasted with the 2°C or greater intervals of Claussen,[7] to obtain definitive measurements on the occurrence of phase transitions suggested by the inflections or "kinks" reviewed by Drost-Hansen. The measurements were made at equilibrium, under an atmosphere of nitrogen with the temperature controlled to 0.01°C, by the drop volume method[14] and the differential capillary rise method.[15] The accuracy of each of the two methods was estimated to be 0.5%. The accuracy of his data and the data of other investigators led Gittens to conclude that the evidence demonstrated the absence of significant "kinks." He, however, said that the subject was not closed but required the establishment of a new measurement method which would provide surface tension measurements to at least 0.001 dyne/cm.

CONVECTION

In lowering the temperature of the water surface by removing heat, the evaporation process raises the surface tension of the water at the surface above the value corresponding to the bulk water beneath the surface.[16] Water in this potentially unstable state tends to exhibit surface tension-driven natural convection. The density of water near the surface is increased by evaporative cooling resulting in instability and a tendency toward gravity-driven convection or "buoyancy-driven" convection.

The name most prominently associated with surface tension-driven flow is Marangoni[17,18] for whom the flow is called the "Marangoni effect." Scriven and Sternling,[19] reviewing Marangoni effects, described them as: (1) motion in an interface caused by local variations of interfacial tension that are "caused in turn by differences in composition or temperature" and (2) "the departure from equilibrium tension that is produced by extension or contraction of an interface, that is by dilational deformation."

Cellular convection in a horizontal fluid layer observed and studied by Bénard[20-24] is familiar to atmospheric dynamicists. Chandrasekhar,[25] according to Berg et al.,[16] established the following fundamental facts concerning natural convection: "First, a certain critical temperature gradient has to be exceeded before stability can set in; second, the motions that ensue on surpassing the critical temperature gradient have a cellular pattern." Berg et al. discussed the experimental methods used for the investigation of convection in horizontal fluid layers: (1) suspended particle methods; (2) optical methods (interferometry, schlieren, and direct shadow); and (3) thermal and other methods. Although the suspended particles furnished a means of detecting the onset of convective stability and a means of observing the flow patterns, the possible effects of the particles on the system make their use questionable. The interferometric method is mentioned elsewhere in this book. The schlieren method will be discussed later in connection with the work of Spangenberg and Rowland.[26]

In the direct-ray method, light rays passing through the medium in a test section are deflected through angles proportional to the refractive index gradient in the medium. The light intensity at a point on a screen corresponding to a point in the test section will be inversely proportional to the divergence of the refractive index gradient at the point in the test section.

Reviews of the optical methods, oriented to wind tunnels, are given by Holder and North[27] and Wood.[28] Berg et al. stated that neither the particle nor optical methods are adequate for quantitative determination of the threshold conditions for convective stability, and described the experiments of Schmidt and Milverton.[29] In these experiments, two horizontal circular brass plates were supported parallel to each other in a body of water in a glass tank. The lower plate was heated electrically. Convection began in the water between the plates when a certain temperature difference between the plates was reached. The separation between the plates was between 4 and 5.5 mm (0.16 and 0.22 in.), and the temperature difference of interest ranged between about 1.5 and 4°C. Plots of temperature difference against the rate of heat transfer exhibited a "break" when instability occurred. Within the limits of experimental error, the temperature difference at which convection began agreed with calculations using a formula developed by Jeffreys[30,31] from theoretical considerations. Using the optical apparatus of Saunders et al.,[32] it was found that the appearance of the image on a screen of light from a point source which had passed through the water also indicated when instability had occurred.

Other thermal methods mentioned by Berg et al.[16] were those of Jarvis[33,34] and Spangenberg and Rowland,[26] both of which are discussed elsewhere in this book. The rapid solidification of a liquid layer undergoing cellular convection for subsequent measurement of surface deformations by Dauzère,[35] and the use by Mysels[36] of the color change in a cobaltous chloride-impregnated filter paper held just above the evaporating surface of water to detect evaporation rates at various areas of the surface were also mentioned.

In discussing comparison of experiment with theory, Berg et al. concluded that further refinement in both experiment and theory was required for quantitative comparisons for analyses of stability for liquid with free surfaces, but that "all the qualitative effects regarding convective stability appear to be in agreement with the predictions of hydrodynamic stability theory." Also, "In summary, it has been demonstrated experimentally that during evaporation several types of convection patterns can exist in quiet shallow (of the order of 1 mm, 0.04 in.) pools of liquid, in addition to the regular hexagons observed by Bénard; the appearance of these flow patterns can be altered drastically by the addition of surface-contaminating molecules."

Spangenberg and Rowland[26] used schlieren photography to study movements induced in water subjected to normal evaporative surface cooling in still air. They used a rectangular plate glass container 10 cm (3.9 in.) in depth filled with tap water and placed in the collimated light beam of a schlieren system. They found that convective currents were established when the estimated Rayleigh number ($Ra \equiv gB\zeta\upsilon^4/k\nu$) was 1193; g is the acceleration due to gravity, B is the coefficient of cubical expansion, ζ is the temperature gradient, ν is the depth of the fluid, k is the thermal diffusivity of the fluid, and ν is the kinematic viscosity of the fluid. Among their conclusions were

1. Schlieren techniques are applicable as remote indicators in the study of movements of transparent liquid where changes of index of refraction are present to serve as the indicator.
2. Orthogonal schlieren photographs showed that when water in a deep container is permitted to cool by natural evaporation, the cooled surface film collects along lines which cause a local thickening of the surface layer and thus establish areas of instability.
3. The observations indicate that the simultaneous presence of both shear and instability are not necessary to form striplike plunging lines in thermal convection as has heretofore been generally assumed.

4. In deep containers with nonlinear temperature gradients, the heat-transfer rate determines the spacing and number of plunging lines. After a particular minimum depth for a given heat-transfer has been exceeded, the number of plunging lines is independent of the container.
5. The surface of the water does not remain plane during convective circulation. It is depressed above the plunging sheets.

Scriven and Sternling[37] examined the onset of steady cellular surface tension gradient-driven convection in a thin layer of fluid in an extension of Pearson's[38] stability analysis. They found a simple criterion for visually distinguishing the dominant force in cellular convection: "in steady cellular convection driven by surface tension, there is an upflow beneath depressions and downflow beneath elevations of the free surface; more accurately, flow is toward the free surface in shallow sections and away in deeper sections. The relationship is just the converse in buoyancy-driven flows, as Jeffreys[39] showed."

Nield[40] found mathematically that the two agencies — surface tension and buoyancy — reinforce one another and are tightly coupled, and that cells formed by surface tension are approximately the same size as those formed by buoyancy.

Since the criteria for the onset of surface tension-driven convection derived theoretically by Pearson[38] and by Sternling and Scriven[41] had not been verified experimentally quantitatively (that is, experimental values of Marangoni numbers at the onset of convection in evaporating layers of pure liquids were consistently much larger than those calculated from Pearson's solution), Vidal and Acrivos[42] undertook to investigate the effect of nonlinear preconvective temperature profiles. The Marangoni number, M, is defined as $M \equiv \psi\zeta\upsilon^2/\mu k$, where ψ is the absolute value of the rate of change of surface tension with temperature, ζ is the temperature gradient, υ is the depth of the fluid, μ is the viscosity of the liquid, and k is the thermal diffusivity of the liquid. The theoretical analysis indicated that the magnitude of the critical Marangoni number could be increased by several orders of magnitude by the existence of nonlinearities in the conductive preconvective temperature profile from that predicted by theory using a linear profile. Experimentally determined values of Marangoni number at the onset of convection in shallow evaporating layers of propyl alcohol were found to be in good agreement with those predicted by the theory.

Effects of Waves

Boyd and Marchello[43] investigated the effects of waves (generated by a series of parallel vertically moving rods in a sealed ripple tank) and cetyl alcohol (n-hexadecanol) on the absorption of CO_2 in water. An effective diffusivity, D_w, was used to correlate the data. Experimental observations led to the conclusion that there was motion of the liquid phase even during no-wave runs. The motion was attributed to natural convection, buoyancy-driven or surface tension-driven (Marangoni) convection. Wave amplitude was considered to be important for mass transfer. For a clean surface, D_w was much larger than the value for molecular diffusivity and decreased with decreasing depth of the water. D_w increased with frequency for the driven runs with a strong dependency on liquid depth, indicating second order drift motion to be an important feature of driven runs. The n-hexadecanol layer greatly reduced D_w. It was concluded that the reduction of D_w by the monolayer in the no-wave case seemed to be the result of curtailment of the Marangoni motion.

Effects of Monomolecular Films

Jarvis,[33] in the introduction to his study of the effect of monomolecular films on the surface temperature and convective motion at the water-air interface, mentioned the following mechanisms through which monomolecular films influence the temperature of a free water surface: (1) the supression of "surface streaming" by adsorbed films of surface-active compounds which are structurally incapable of reducing evaporation;[70] (2) the retarding of convective cellular motion; (3) the reduction, in proportion to the surface viscosity of the monomolecular film, of surface motion generated by a flow of air over the interface; and (4) the exertion of a viscous drag on the water immediately under the film. Jarvis made an experimental study of the effects of these mechanisms and of the air temperature, air relative humidity, and the air velocity over the surface on surface temperature to determine under what conditions an adsorbed monomolecular film might be expected to alter the temperature of a clean water surface. The experiments were performed in a glass tank 45 cm (18 in.) long, 25 cm (9.8 in.) wide, and 23 cm (9.1 in.) deep with a Lucite cover; a stream of air or other gas could be introduced through a 0.5-in. internal diameter Lucite tube mounted in the lid. Three thermistors (each mounted directly below the gas inlet) in a waterproof glass probe with a maximum diameter of 0.2 cm (0.08 in.) and an approximately length of 1 cm (0.4 in.) were mounted one each at depths of 15 cm (5.9 in.) and 3 cm (1.2 in.) and one near the surface. The average depth of water over the near-surface thermistor was 2 mm (0.08 in.), easily varied from 0 to 3 mm (0.12 in.). Temperature changes of $\pm 0.01°C$ could be detected and recorded. The relative humidity could be varied from about 0 to 95%, and the gas flow rate could be varied from 0 to 10 L/min. The water used in the tank was from a tin-lined still and was allowed to equilibrate with CO_2 in the atmosphere. Movement of the surface water was followed by observing the motion of talc particles; movement of the bulk water was monitored by following the motion of finely divided charcoal particles with a telescope.

Comments of particular interest are

1. "In the presence of a monolayer, a thin layer of surface water will be unable to enter into the convective exchange of the surface and bulk water and will be a factor influencing the surface temperature."
2. "The surface temperature of a body of water confined in a glass vessel, where the heat transfer through the glass is small, will be largely determined by: (1) the rate at which heat is lost by evaporation, (2) the rate of radiative heat exchange with the atmosphere, and (3) the convective or turbulent heat transfer of warmer water to the surface layer."
3. The rate of evaporation will be primarily a function of the relative humidity of the air (assuming the air temperature and bulk water temperature remain constant and assuming a low rate of air flow).
4. Convective or turbulent movement of water in a clean water surface is affected by the rate of air flow and by changes in evaporation or relative humidity, e.g., low relative humidity and rather low flow rates lead to convective or turbulent movement in which cool and warm water move alternately past the near-surface thermistor.

Among the conclusions of Jarvis were

1. "An adsorbed monomolecular film may influence surface temperature not only by retarding evaporation but also by having an effect on the convective or turbulent motion already present in the water surface."
2. The localized convective exchange of cooler, more dense surface water for warmer water beneath (where evaporation is proceeding rapidly at low rates of air flow) is reflected in rapid temperature fluctuations recorded by the near-surface thermistor.
3. A monolayer of oleic acid (which does not reduce the evaporation) on the surface (under conditions of item 2) did not change the surface temperature or the water movement pattern, monolayers of cetyl alcohol or stearic acid (which do retard evaporation) increased surface water temperature.
4. At the higher flow rates a rapid transport of bulk water into the interface was observed, the addition of a monolayer effectively stopped convection and permitted an immobilized thin layer of surface water to cool (due to evaporation) as much as 0.4°C (the bulk water temperature was 19.20 ± 0.05°C, and the temperature of the gas was 20.5 ± 0.2°C).
5. Any monolayer with sufficiently high surface viscosity will be capable of influencing surface temperature (in the manner of item 4), due to viscous drag exerted by the monolayer on the underlying water.
6. "By the same method surface contamination can greatly alter the rate of evaporation of organic liquids, particularly under vacuum, by reducing convection and surface streaming."

Navon,[44] in investigating the effects of insoluble monomolecular surface films on the evaporation of water and on the transfer of heat from the water to the surface, used the reversal in the temperature dependence of the density of water at 4°C to determine the role of natural convection in evaporative heat transfer. A substantial increase in the Nusselt number, Nu, ($Nu \equiv \Theta_h \upsilon / k$, where Θ_h is the heat transfer coefficient, υ is the thickness of the liquid layer, and k is the thermal conductivity) the ratio of total heat transfer to that due to conduction alone, was observed for clean water surface at bulk water temperature higher than 4°C. This increase indicated that evaporative cooling of the surface layer induced gravity-driven convection which enhanced both heat transfer to the surface and the evaporation rate.

Effect of Uniform Rotation

Vidal and Acrivos[45] investigated, both theoretically and experimentally, the effect of uniform rotation on surface tension-driven convection in an evaporating fluid layer. In introducing the investigation, they made the following comments which are of general interest: the presence of convective flows increases the rate of transport of energy and matter; in the determination of certain physical properties such as the condensation coefficient or the interfacial resistance to mass transfer from a liquid to a gas phase it is imperative to suppress convective currents; buoyancy-driven convection can be eliminated by operating with relatively shallow liquid layers (less than about 5 mm, 0.2 in., deep); surface tension-driven convection seems to persist even in pools 0.5 mm (0.02 in.) in depth; and although this type of convection can be eliminated, especially for water by using small amounts of surface active agents (Berg and Acrivos[46]), such a technique greatly affects the nature of the surface.

In view of the extraordinary stability of rotating fluids, it appeared to Vidal and Acrivos to be logical to assume that the three-dimensional convective motions in evaporating layers could be damped by subjecting the system to uniform rotation. Chandrasekhar[47] had already demonstrated this fact for the buoyancy-driven case. Therefore, it was the purpose of the Vidal and Acrivos paper to investigate theoretically the possible influence of such a Coriolis force on surface tension-driven convection and to correlate the results of the mathematical analysis with the results of an experimental study of convection in uniformly rotating evaporating systems. The theoretical analysis followed the small-disturbance approach of perturbation and led, at the neutral state, to a functional relation between the Marangoni and Taylor numbers which was then computed numerically.

The Marangoni number was defined earlier. The Taylor number, T, is defined as $T \equiv \Omega^2 \upsilon^4 / \nu^2$, where Ω is the speed of rotation, ν is the kinematic viscosity of the fluid, and υ is the thickness of the liquid layer. It was shown analytically that, in the limit of rapid rotation, the velocity and temperature fluctuations are confined to a thin Ekman layer near the surface of the evaporating fluid and that the critical Marangoni number for neutral stability is $M_c = 4.42\ T^{1/2}$ and that the critical wave number is $\alpha_c = 0.5\ T^{1/4}$.

The experimental study in the investigation of Vidal and Acrivos was made using a schlieren system described by Berg et al.[48] The liquid, one of a variety of pure organic liquids and mixtures (in the major part of the experiment, a 50% solution by volume of ethyl ether in n-heptane was used), was placed in a circular glass dish of 6-in. diameter set on a glass turntable that could be rotated at speeds up to 1/2 revolution/sec. The meniscus height was kept at a value less than about 15% of the average depth of the pool, and circular waves caused by unavoidable vibrations in the system were eliminated by introducting a circular beach having a slope of 30° and covering the outer half of the bottom of the dish. For the ethyl ether-n-heptane solution, the evaporation of ethyl ether (the more volatile component) resulted in a potentially unstable configuration with respect to the surface tension mechanism alone since the ethyl ether has a lower surface tension and a higher density than n-heptane.

Berg et al.[48] had reported that the convective pattern of evaporating shallow layers (of depth less than 0.5 cm, 0.2 in.) had a cellular structure which, upon increase in depth (greater than 1.0 cm, 0.40 in.), became highly irregular and nonstationary. From the mathematical analysis of Vidal and Acrivos it would be expected that the stabilizing effect of the Coriolis force in a moderately deep layer of rotating evaporating liquid would bring the system closer to its marginal state and that above a certain value of the Taylor number a cellular convective pattern would be recovered. In all liquids studied, the change from random motion to a regular flow structure with increasing T, T greater than or equal to 1000, was observed.

Glass microballoons were suspended in the fluid, and their motions were followed to determine how far below the surface the convective motion propagated. It was found that for T greater than 1000 the three-dimensional motions were confined to a small portion of the liquid adjacent to the surface, in contrast with penetration of convective currents down to the bottom of the layer when the evaporating pool was not rotating. The convective sublayer became thinner as the speed of rotation was increased. Among the conclusions were (1) uniform rotation enhances the stability of the system; (2) the marginal state of the system is stationary; (3) for large values of T the dominant convective pattern consists of hexagonal cells that diminish in size as T increases; and (4) in the presence of uniform rotation, the direction of flow is upward in the core and downward along the cell boundaries when convection is surface tension-driven, whereas the reverse holds whenever the convection is driven by buoyancy and the system is cooled from above.

Effects of a Progressive Wave

O'Brien and Omholt[49] predicted surface temperature and heat flux variations induced by a progressive wave on a thin layer of water across which steep thermal gradients exist. Noting that it is generally recognized[50-52] that a thin thermal layer, across which there is an appreciable gradient of temperature, exists at the surface of the ocean, they stated that in a typical intense manifestation the layer is of the order of 1 mm (0.04 in.) in depth and that the temperature difference is of the order of 1°C. They investigated the hypothesis that only molecular conduction and laminar convection (excluding buoyancy-induced convection) contribute to the heat transfer within the thermal layer when the interface is a progressive wave and concluded that these two mechanisms do not account for the behavior of surface temperature observed in a wave tank.[53]

Liquid Flow Patterns

Leontiev and Kirdyashkin[54] made an experimental study of liquid flow patterns in a horizontal layer near a horizontal surface of natural convection and measured the temperature profile in an individual cell and showed that an eddy cell flow existed near the heat transfer surface.

Heat Transfer and Thermal Structure in a Boundary Layer

Katsaros and Garrett[55] used a convection tank to study heat transfer and thermal structure in the boundary layer below an air-water interface with the addition of monomolecular films at the surface. The results indicated that buoyancy-driven fluctuations are so great as to dominate over those introduced by a 1 m/sec wind, and that buoyancy-driven motions mitigate the effects of surfactants. They conclude that: "The implication then is that the internal dynamics of a water body and the external conditions in the atmosphere above are as important as the chemical properties of a surfactant in affecting the evaporation rate."

Evaporation Regimes in Heated Ponds

Adams et al.[56] provided theoretical analyses for each of four evaporation regimes that could be experienced at heated ponds, ranging from free convective layers to forced convective layers. Free convection is that due to the horizontal movement of air across the water surface. Forced convection is that caused by the density difference between air at the water surface and in the ambient surroundings. Predictive equations for evaporation under a range of conditions from free convection to forced convection were reexamined. A new equation was developed that combines free and forced convection in a different manner than from previously developed equations. Evaporation was computed as the square root of the sum of the squares of the components.

Flow Patterns in an Evaporating Liquid Layer

Berg et al.[48] applied schlieren optics to reveal a variety of flow patterns in an evaporating liquid layer; to obtain qualitative information on both the structure and scale of the flows; and particularly the manner in which they are affected by fluid properties, fluid depth, and the presence of surface contamination.

Experiments were performed with water, acetone, benzene, carbon tetrachloride, *n*-heptane, isopropanol, methanol, and a number of binary solutions. The materials were evaporated into still air from liquid layers 10 × 10 cm (3.9 × 3.9 in.) in horizontal extent, with depths ranging from 0.25 to 10 mm (0.0098 to 0.39 in.).

The schlieren technique permits one to see and to photograph gradients of refractive index, and hence gradients of temperature and composition, in a transparent medium. The technique can be used to visualize flow patterns in such systems, since an evaporating fluid will have generally a nonuniform temperature distribution owing to the presence of convective flow.

Water did not behave like the other fluids. No convection was evident until the depth reached about 1 cm. At that depth and beyond, the pattern was one of cold streamers along. This "anomalous" behavior of water might have been due to surface contamination by surface-active agents. Two experimental observations tended to collaborate this possibility. The presence of small amounts of surface contamination had a profound effect on the morphology of the developed convection in those cases in which the surface tension mechanism played an important role. It was possible to identify certain flow patterns as being induced by a surface tension-driven instability and others as being due to the buoyancy-driven convection.

WIND GENERATED WAVES

Hidy and Plate,[57] in the introduction to a paper on the action of wind on water standing in a laboratory channel wrote the following paragraph which is quoted directly since it serves as an introduction to the subject of wind generated waves:

> When turbulent air passes over water initially standing in a channel, small surface waves grow with fetch and duration of air motion, the aerodynamic roughness of the water changes, causing modification in the vertical profile of the mean air flow over the water surface. Energy is transferred from the air motion to the water and is partitioned between the drift current and the surface waves. The wind waves in turn alter the air velocity distribution. Thus, a complex mechanism of air-water interaction and feedback takes place which as yet is not fully understood in all its details.

The paper was intended, in reporting new measurements of combined air and water motion in a laboratory channel, to contribute toward filling the gap between theory and experiment. The experiments were performed in a wind-water tunnel, described by Plate,[58] consisting of a smooth-bottomed wind tunnel 0.61 m (2.0 ft) wide, 0.76 m (2.5 ft) high, with a plexiglass test section about 12 m (39 ft) long. Water depths of up to 15 cm (5.9 in.) could be maintained during operation. Sloping beaches of aluminum honeycomb, shaped in such a way that there was a smooth transition between the air and water flow, prevented reflection of waves at the inlet and outlet. An axial fan controlled the air motion, made uniform by fine mesh screens and honeycombs, through the tunnel. The air speed was measured by a Pitot-static tube in conjunction with an electronic micromanometer. The probe could be positioned anywhere in a given section from the bottom of the tunnel to about 10 cm (3.9 in.) from the top. Pressure gradients in the air and the depth of water were measured by piezometer taps, spaced at 1.2-m (3.9-ft) intervals down the tunnel, connected to glass tube manometers.

Phase speeds and wave lengths of significant waves were determined from motion pictures. Although it was not possible to determine velocity distribution in the water, the surface velocity was estimated by measuring the time of transit of a buoyant particle on the water between fixed stations. The displacement of the water surface as a function of time was measured by capacitance probes, consisting of 34-gauge copper magnet wires stretched vertically across the tunnel perpendicular to the water surface. The copper wires and the water formed two plates of a capacitor, the Nyclad wire insulation serving as the dielectric. The instrumentation associated with the probes provided a recorded amplitude linear with the varying water depth with a flat (1%) response to about 30 Hz. Air velocities taken 20 cm (7.9 in.) above the surface of the water that was initially standing in the channel were varied from 0 to 15 m/sec (49 ft/sec) while the depth of the water was changed from 2.5 to 10 cm (0.98 to 3.9 in.). Under steady flow conditions, the properties of the fluid motion were observed at fetches of about 1.8 to 12 m (5.9 to 39 ft).

The conclusions of Hidy and Plate include the following: (1) many features of the air-water flow could be described in terms of a two-dimensional picture; (2) the wind-generated waves traveled downstream at approximately the same phase speed as gravity waves of small amplitude provided that the effects of drift current were taken into account; (3) the average drag coefficient (wind on water) increased with increasing wind speed; (4) the aerodynamic roughness of the water surface and the friction velocity were related to the height of the waves; (5) the wind-generated waves consisted (visually) of nearly periodic waves on which random wavelets were superimposed; and (6) the wave development could be explained reasonably well by a mechanism by which the growth of waves is initiated by turbulent pressure fluctuations in the air flow over the water, and is augmented by energy transfer from the air to the water through shearing flow instability.

Plate et al.,[59] stating that the results of Hidy and Plate could not be conclusive concerning testing of the Phillips-Miles theory since not all the conditions required by the theory were met in the experiments, undertook more accurate experiments to clarify features of the mechanism of wave generation by wind. They described the Phillips-Miles mechanism in the following way:

> According to Phillips[60] the initial wavelets derive from the resonance between the free modes of water surface and the pressure field of the turbulent motion in the air flow. This class of water waves is predicted to grow linearly with time (or with fetch) until the viscous mechanism of Miles[61] takes over. Then waves and air flow become coupled; the waves give rise to a pressure distribution at the water surface which feeds energy to the waves at a rate which exceeds the energy gain due to the induced pressure; single components of the wave spectrum grow exponentially with time or fetch under the action of a steady wind.

Two types of experiments were performed. In the first series, the development of the waves from their inception to fully developed waves for two different air speeds was observed. The capacitance gauge, described before for the Hidy and Plate paper, capable of detecting water surface undulations of less than 10 μm (3.9×10^{-4} in.) in amplitude was used to determine the variation in the water surface elevation. Turbulent intensities and spectra were determined from single hot wire anemometer measurements made in the air flow under similar conditions.

The second experiment was designed to duplicate as nearly as possible the conditions required by the Miles model. The experiments were performed in the wind-water tunnel described in connection with the Hidy and Plate paper. Air flowed from the tunnel over a smoothly sanded aluminum plate of 3.7 m (12 ft) length onto a body of initially standing water, 11.4 cm (4.5 in.) deep, and 13.7 m (45 ft) long. An aerodynamically smooth flow in a developing turbulent boundary layer was assured by screens and a honeycomb section placed at the inlet of the tunnel. Measurements were made of water surface displacement, profiles of mean air velocity, turbulent fluctuations of the longitudinal components of the air flow, apparent critical wind speeds, and velocities of the drift current at the water surface. The measurements of turbulent fluctuations were made with a Hubbard-type hot wire anemometer working on the constant temperature principle. At a given fetch the critical wind speed was defined as the air speed at a fixed location at which the vibration of the water became visible, as detected by observing the reflection of the capacitance wire in the water surface. The drift current velocity was measured by timing the passage of thin wax paper discs, 0.6 cm (0.24 in.) in diameter, between markers.

Among the conclusions of Plate et al. were the following: (1) there was no indication that the first waves were produced by direct interaction of the water surface with the air turbulence and no significant feedback of the wave into the turbulence could be detected; (2) under the shearing action of the wind the first waves grew exponentially, the growth rates agreeing to a fractional error of 61% or less with estimates from the viscous shearing mechanism of Miles; (3) a slight, but not significant, improvement was obtained with the theory of Drake in which the Miles model is extended to include the effect of drift current induced by the wind; and (4) the experiments were sufficiently accurate that discrepancies between experiments and theory must be attributed to theory or its interpretation.

Hess et al.[62] reported some new preliminary experimental results on the influence of the perturbing action of waves on the air motion overhead. The experiments were performed in the wind-water tunnel described before. Hess et al. concluded that the data for the distribution of turbulent intensities above a wavy water surface are qualitatively consistent with boundary layer flow over smooth or rough solid surfaces, that only the Reynolds sheering stress showed an anomalous decrease with z/δ (the ratio of wave height to boundary layer thickness) which they thought was evidently associated with the waves, and that the fact that no strong wave component was found in the turbulence spectrum provided indirect evidence for separation and thus tended to confirm Schooley's[63] visual observation of separation.

Effects of Waves on Evaporation

Easterbrook[64] used a wave tank to study the effects of waves on evaporation from a free-water surface. The tank was a large closed trough, 40 ft long, 4 ft wide, and 3 ft deep. A paddle at one end of the tank produced waves which traveled down the trough and were dissipated on a gently sloping "beach." A fan circulated air over the water surface. A dead air space over the water at the beach end of the tank reduced the variable evaporation effects due to the breaking of the waves. Temperature and humidity measuring instruments were mounted in the return duct of the closed system; small thermistors were used for measuring air temperature, a hot wire anemometer was used for wind measurements about midway between the surface of the water and the top of the main duct. A mercury thermometer was mounted about 2 in. above the bottom of the tank for measuring bulk water temperature.

The results for the wave-tank experiments showed a definite relationship between surface wave conditions and measured evaporation rate. It appeared that certain combinations of wind speed and wave conditions led to evaporation rates smaller than those measured with similar wind speed and smaller waves or smaller ratio of wave height to wave period. Easterbrook concluded that the buildup of high humidity in vortices and adjacent dead air regions, detected in the airflow over the waves and apparently trapped in the wave troughs and moving along with the wave system, tended to reduce the transfer of water across the air-water boundary; and also, that the reduced turbulent mixing between the dead air spaces and the upper flow inhibited vertical transport of moisture. A field study of evaporation at Lake Hefner, Oklahoma was undertaken to provide the information to connect the tank studies, in which the air flow patterns were thought to be modified by the confined nature of the evaporation duct, to the field situation.

Massaldi et al.[65] studied the effect of interfacial waves on the rate of evaporation of water from a free-water surface. For cases in which the resistance to evaporation was located in the liquid phase, it had been generally agreed that the rate of heat transfer is greatly enhanced by wave motion. Situations in which the gas phase controls the resistance had not been studied as much.

For the experimental work, a recirculating wind tunnel (in which the air velocity was kept low) was used. Waves were generated under conditions that ensured there would be no atomization. A rectangular pan — 22 cm (8.7 in.) wide, 2.5 cm (0.98 in.) deep, and the same length as the test section of the wind tunnel, 95 cm (37 in.) — was completely filled with water so that the surface of the water was even with the lower wall of the tunnel.

Dry- and wet-bulb temperatures were measured by thermometers placed in the bulk of the airstream. The air velocity was measured by an accumulating anemometer. The liquid level in the pan was measured using a contact level meter provided with a vernier in a tube placed outside the tunnel but communicated to the pan.

An evaporation measurement run was started by measuring the initial water level and noting the time. The level was followed each half hour until there was a total decrease in level of 1.5 to 2 mm (0.06 to 0.08 in.). The rate of evaporation was calculated from the initial and final level readings, and the time. The evaporation rate increased in the presence of waves. The increase was ascribed to two effects: (1) increase in the transfer area and (2) induction of turbulence in the gaseous boundary layer. The second effect was approxiamtely four times as large as the first. The results were in qualitative agreement with results reported in the literature concerning the increase of friction factor in the presence of interfacial waves. The results were within the order of magnitude that could be expected for waves of 1 mm (0.04 in.) amplitude.

The presence of interfacial waves increased substantially the gas mass transfer coefficient between gas-liquid phases relative to that for a stagnant interface, mainly through the induction or turbulence in the gas boundary layer.

Dick and Marchello[66] conducted a small-scale laboratory study of the effect of capillary-gravity waves on evaporation. Evaporation was carried out in a system of co-currently flowing air and water. Distilled water evaporated from a 50.8 × 15.9 × 10.2 cm (20 × 6.25 × 4.0 in.) plexiglass tank having a wave generator situated in the center.

Humidities of airstreams were measured using an electric hygrometer. Temperatures were measured by five-member thermopiles. Before each run the distilled water was aerated and agitated for 2 hr. After temperatures and humidities for the clean water had been recorded, a solution of hexadecanol was introduced by a syringe through ports in the cover of the tank. Hexadecanol-petroleum ether in the amount of 1 cm^3 corresponded to 26.56 $\times 10^{-10}$ mol/cm^2 on the surface. Temperatures and humidities were then monitored as before.

After the data for no-wave runs were obtained, the wave generator was turned on and the evaporation parameters measured. The study was concerned with measuring the effectiveness of a surfactant in reducing evaporation from a free-water surface. The mass transfer coefficient obtained from the data lent itself to a comparison of the evaporation with and without the film.

The mass transfer coefficients for three runs showed that small amplitude waves had little effect on evaporation. The waves destroyed much of the retarding properties of the film.

Trapeznikov[67] found through direct measurements that a small wave weakens the capacity of a monolayer for reducing the evaporation rate of water, due to the weaker cohesion in the monolayer. This was confirmed by the influence of a wave on the viscosity of a monolayer and a surface layer in a solution and in a two-sided film.

Healy and La Mer[68] introduced a mechanical vibration, waves without wind, on a water surface initially covered by a monolayer and monitored the properties of the monolayer subjected to this disturbance. The vibration created capillary waves, waves with wavelength of less than 1.0 cm (0.4 in.).

The evaporimeter used consisted of a Langmuir trough, Wilhelmy plate assembly, desiccant box assembly to record the evaporation resistance of the monolayer, and wave generator. The wave generator was a T-bar lying in the water surface, vibrating in and out with respect to the water surface. The monolayer materials were high purity samples of straight-chain alcohols of 14 to 20 carbons. The spreading solution was n-hexane. Surface pressure-area per molecule isotherms were measured while the disturbance was present.

Attempts to measure evaporation resistance while the surface was disturbed were only partially successful. The capillary waves, at surface pressures of the order of 30 dyne/cm, reduced the evaporation resistance by about 25 to 30%. The grinder-duster technique,[69] in which solid flakes of monolayer-forming material are present in the monolayer, is ideally suited to provide immediate repair of sections of the monolayer damaged by wave action. For high-amplitude capillary waves, the surface pressure behavior at pressures less than about 15 dyne/cm could be understood in terms of the increase in area due to the wave. At higher pressures, this increase in area was not sufficient to explain the isotherm for the wave-covered surface.

REFERENCES

1. Drost-Hansen, W. "Aqueous Interfaces. Methods of Study and Structural Properties," Parts I and II, in *Chemistry and Physics of Interfaces* (Washington, DC: American Chemical Society, 1965).
2. Addison, C. C. "The Properties of Freshly Formed Surfaces. Part I. The Application of the Vibrating Jet Technique to Surface Tension Measurements on Mobile Liquids," *J. Chem. Soc. (London) Proc.* 535 (1943).
3. Addison, C. C. *J. Chem. Soc. (London) Proc.* 354 (1945).
4. Pallasch, R. "Dynamic and Static Surface Tension of Liquid Bell," *Ann. Phys.* 40:463 (1941).
5. Jamieson, D. T. "Condensation Coefficient of Water," *Nature* 202:583 (1964).
6. Göring, W. "Dependence of Surface Tension Upon the Rates of Formation and the Ageing of the Surface," *Z. Electrochem.* 63:1069 (1959).
7. Claussen, W. F. "Surface Tension and Surface Structure of Water," *Science* 156:1226 (1967).
8. Moser, H. "The Absolute Value of the Surface Tension of Pure Water by the Loop Method and its Dependence on Temperature," *Ann. Phys.* 82:993 (1927).
9. Eötvös, R. "Concerning the Correlation of Surface Tension of Liquids with Their Molecular Volume," *Ann. Phys.* 27:448 (1886).
10. Ramsay, W., and J. Shields. "The Variation of Molecular Surface Energy with Temperature," *Philos. Trans. R. Soc. (London)* A184:647 (1893).
11. Teitel'baum, B. Ya, T. A. Gortolova, and E. E. Sidorova. "Surface Tension at Various Temperatures of Aqueous Solutions of Lower Alcohols," *Zh. Fiz. Khim.* 25:911 (1951).
12. Grastella, J. "Superficial Structure of Water and Surface Phenomena," *J. Chem. Phys.* 44:306 (1947).
13. Gittens, G. J. "Variation of Surface Tension of Water with Temperature," *J. Colloid Interface Sci.* 30:406 (1969).
14. Gaddum, J. H. *Proc. R. Soc. (London)* B109:114 (1931).
15. Harkins, W. D., and A. E. Alexander. *Physical Methods of Organic Chemistry*, 3rd ed., A. Weissberger, Ed., (New York: Interscience Press, 1959), p. 757.
16. Berg, J. C., A. Acrivos, and M. Boudart. "Evaporative Convection," *Adv. Chem. Eng.* 6:61 (1966).
17. Marangoni, C. *Nuovo Cimento,* Ser. 2, 16:239 (1871).
18. Marangoni, C. *Nuovo Cimento,* Ser. 3, 3:97 (1878).
19. Scriven, L. E., and C. V. Sternling. "The Marangoni Effects," *Nature* 187:186 (1960).
20. Bénard, H. "Cellular Eddies in a Liquid Surface Transporting Heat by Steady State Convection," *Ann. Chim. Phys.*, Ser. 7, 23:62 (1901).
21. Bénard, H. "On the Formation of Lunar Rings, From the Experiments of C. Dauzère," *Comp. Rend.* 154:260 (1912).
22. Bénard, H. "On Cellular Eddies and the Theory of Rayleigh," *Comp. Rend.* 185:1109 (1927).
23. Bénard, H. "On Eddies in Bands and the Theory of Rayleigh," *Comp. Rend.* 185:1257 (1927).

24. Bénard, H. "Is the Surface Solar Photosphere a Layer of Cellular Eddies," *Comp. Rend.* 201:1328 (1935).
25. Chandrasekhar, S. "Thermal Convection," *Daedalus* 86:323 (1957).
26. Spangenberg, W. B., and W. R. Rowland. "Convective Circulation in Water Induced by Evaporative Cooling," *Phys. Fluids* 4:743 (1961).
27. Holder, D. W., and R. J. North. "Optical Methods for Examining the Flow in High-Speed Wind Tunnel, Part 1, Schlieren Methods," *Agardograph* 23, (November 1956).
28. Wood, G. P. "Optical Methods for Examining the Flow in High-Speed Wind Tunnel, Part II, Interferometric Methods," *Agardograph* 23 (November 1956).
29. Schmidt, R. J., and S. W. Milverton. "On the Instability of a Fluid When Heated from Below," *Proc. R. Soc. (London)* A152:586 (1935).
30. Jeffreys, H. "The Stability of a Layer of Fluid Heated Below," *Philos. Mag.*, Ser. 7, 2:833 (1926).
31. Jeffreys, H. "Some Cases of Instability in Fluid Motion," *Proc. R. Soc. (London)* A118:195 (1928).
32. Saunders, O. A., H. Fishenden, and H. D. Mansion. "Some Measurements of Convection by an Optical Method," *Engineering* 139:483 (1935).
33. Jarvis, N. L. "The Effect of Monomolecular Films on Surface Temperature and Convective Motion at the Water/Air Interface." *J. Colloid Sci.* 17:512 (1962).
34. Jarvis, N. L., C. O. Timmons, and W. Zisman. *Retardation of Evaporation of Monolayers*, V. K. LaMer, Ed., (New York: Academic Press, 1962), p. 41.
35. Dauzère, C. "Investigations on Solidification," *J. Phys.*, Ser. 4, 6:892 (1907).
36. Mysels, K. J. "Direct Observation of Evaporation from Quiescent Water," *Science* 129:96 (1959).
37. Scriven, L. E., and C. V. Sternling. "On Cellular Convection Driven by Surface-Tension Gradients: Effects of Mean Surface Tension," *J. Fluid Mech.* 4:489 (1958).
38. Pearson, J. R. A. "On Convection Cells Induced by Surface Tension," *J. Fluid Mech.* 4:489 (1958).
39. Jeffreys, H. "The Surface Elevation in Cellular Convection," *Quart. J. Mech. Appl. Math.* 4:283 (1951).
40. Nield, D. A. "Surface Tension and Buoyancy Effects in Cellular Convection," *J. Fluid Mech.* 19:341 (1964).
41. Sternling, C. V., and L. E. Scriven. "Interfacial Turbulence: Hydrodynamic Instability and the Marangoni Effect," *AIChE J.* 5:514 (1959).
42. Vidal, A., and A. Acrivos. "Effect of Nonlinear Temperature Profiles on the Onset of Convection by Surface Tension Gradients," *Ind. Eng. Chem. Fundam.* 7:53 (1968).
43. Boyd, D. P., and J. M. Marchello. "Role of Films and Waves on Gas Absorption," *Chem. Eng. Sci.* 21:769 (1966).
44. Navon, U. "Effects of Monomolecular Films on Evaporational Heat and Mass Transfer of Water," Department of Aerospace and Mechanical Sciences, Princeton University, Final Report under Grant 14-01-0001-438, Office of Saline Water, U.S. Dept. of Interior (October 1967).

45. Vidal, A., and A. Acrivos. "The Influence of Coriolis Force on Surface-Tension-Driven Convection," *J. Fluid Mech.* 26:807 (1966).
46. Berg, J. C., and A. Acrivos. "The Effect of Surface Active Agents on Convection Cells Induced by Surface Tension," *Chem. Eng. Sci.* 20:737 (1965).
47. Chandrasekhar, S. *Hydrodynamics and Hydrodynamic Stability* (Oxford Clarendon Press, 1961).
48. Berg, J. C., M. Boudart, and A. Acrivos. "Natural Convection in Pools of Evaporating Liquids," *J. Fluid Mech.* 24:721 (1966).
49. O'Brien, E. E., and T. Omholt. "Heat Flux and Temperature Variation at a Wavy Water-Air Interface," *J. Geophys. Res.* 74:3384 (1969).
50. Ewing, G., and E. D. McAlister. "On the Thermal Boundary Layer of the Ocean," *Science* 131:374 (1960).
51. Hasse, L. "On the Cooling of the Sea Surface by Evaporation and Heat Exchange," *Tellus* 15:363 (1963).
52. Saunders, P. M. "Aerial Measurement of Sea Surface Temperature in the Infrared," *J. Geophys. Res.* 72:4109 (1967).
53. van de Watering, W. P. M., and D. C. Wiggart. "Surface Temperature Fluctuations Due to Waves," *Trans., Am. Geophys. Union* 49:204 (1968).
54. Leontiev, A. I., and A. G. Kirdyashkin. "Experimental Study of Flow Patterns and Temperature Fields in Horizontal Free Convection Liquid Layer," *Int. J. Heat Mass Transfer* 11:1461 (1966).
55. Katsaros, K. B., and W. D. Garrett. "Effects of Organic Surface Films on Evaporation and Thermal Structure of Water in Free and Forced Convection," *Int. J. Heat Mass Transfer* 25:1662 (1982).
56. Adams, E. E., D. J. Cosler, and K. R. Helfrich. "Evaporation from Heated Water Bodies: Predicting Combined Forced Plus Free Convection," *Water Resour. Res.* 26:425 (1990).
57. Hidy, G. M., and E. J. Plate. "Wind Action on Water Standing in a Laboratory Channel," *J. Fluid Mech.* 26:651 (1966).
58. Plate, E. J. "A Research Facility with Concurrent Air and Water Flows," *La Houille Blanche* 6:595 (1965).
59. Plate, E. J., P. C. Chang, and G. M. Hidy. "Experiments on the Generation of Small Water Waves by Wind," *J. Fluid Mech.* 35:625 (1969).
60. Phillips, O. M. "On the Generation of Waves by Turbulent Wind," *J. Fluid Mech.* 2:417 (1957).
61. Miles, J. W. "On the Generation of Surface Waves by Shear Flows, Part 4," *J. Fluid Mech.* 13:433 (1962).
62. Hess, G. D., G. M. Hidy, and E. J. Plate. "Structure of Turbulence in Air Flowing over Small Water Waves," Submitted to *Phys. Fluids* (1969).
63. Schooley, A. H. "Simple Tools for Measuring Wind Fields above Wind-Generated Water Waves," *J. Geophys. Res.* 68:5497 (1963).
64. Easterbrook, C. C. "A Study of the Effects of Waves on Evaporation from Free Water Surface," Cornell Aeronautical Laboratory Report No. RM-2151-P-1, April 15, 1968.
65. Massaldi, H. A., J. C. Gottifredi, and J. J. Ronco. "Effect of Interfacial Waves on Mass Transfer During Evaporation of Water from a Free Surface," *Lat. Am. J. Chem. Eng. Appl. Chem.* 6:161—170 (1976).

66. Dick, G. F., Jr., and J. M. Marchello. "Effect of Small Surface Waves on Evaporation through Monolayers," *Water Resour. Res.* 5:395—400 (1969).
67. Trapeznikov, A. A. "Rheological Properties of Two-Sided Films and Monolayers and the Influence of the Latter on Evaporation of Water Damping of Small Waves," *Chim. Phys. Appl. Prat. Aq. Surf.* 2:95—108 (1969).
68. Healy, T. W., and V. K. La Mer. "The Effect of Mechanically Produced Waves on the Properties of Monomolecular Layers," *J. Phys. Chem.* 68:3535—3539 (1964).
69. Vines, R. G., in *Retardation of Evaporation* (New York: Academic Press, 1962) V. K. La Mer, Ed., p. 150.
70. Ewing, G. C., private communication.

CHAPTER 6

Surface Temperature, Temperature Differences between Surface and Bulk, Temperature Gradients, and Humidity Gradients above a Water Surface

INTRODUCTION

It is apparent that the water surface temperature is a quantity of primary importance in the evaporation of water. It is therefore appropriate to inquire how well temperature measurements can be made at the liquid water surface. In this regard it is of interest to estimate the effect of the inaccuracy of the temperature measurement on the calculated rate of evaporation and on the calculated evaporation coefficient.

In the vicinity of 20°C, an error of 1°C in the determination of the water surface temperature results in an error of about 6% in saturation water vapor pressure, a similar error in the calculated value of the evaporation coefficient, $\bar{E}$; and with known $\bar{E}$ an error of about 6% in calculated evaporation rate. These remarks apply to evaporation into vacuum. It is appropriate to review selected accounts of pertinent water surface temperature measurements in the literature.

MEASUREMENT OF SURFACE TEMPERATURE

Pasquill[1] used a "surface thermometer" of the Negretti and Zambra type, consisting of a mercury-in-glass thermometer with a thin metal disk attached to the thermometer bulb, to approach the measurement of surface temperature.

Yamamoto[2] used a copper (0.234-mm, 0.009-in., diameter wire)-constantan (0.254-mm, 0.001-in., diameter wire) thermocouple to measure the water surface temperature.

To measure the surface temperature of evaporating propyl alcohol, Vidal and Acrivos[3] used an iron-constantan thermocouple junction made by fusing the tips of two thin Teflon®*-coated wires; the junction had an overall diameter of 0.2 mm (0.008 in.). The wires were coated with silicone cement, leaving only the fused junction exposed to minimize the evaporative cooling of the wires. The thermocouple was brought into contact with the liquid surface, and a wooden lid having a narrow slit to accommodate the thermocouple was

* Registered trademark of E. I. du Pont de Nemours and Company, Inc., Wilmington, Delaware.

used to cover the apparatus until equilibrium had been attained. The lid was then carefully removed, and the output from the thermocouple was continuously recorded.

Hickman described and simulated the method of Baranev[4,5] for determining the temperature of an evaporating surface. According to Hickman's description of Baranev's method:

> A thin walled test tube was immersed in a heating bath and connected to a vapor removal system. The tube was filled with test liquid to well above a delicate thermocouple, after which pump and timer were started. The liquid began to evaporate and the level receded at a noted pace and eventually overtook the thermocouple, the temperature of which had been steadily falling. As the couple and "surface" coincided, the temperature steadied for as long as the couple could maintain a meniscus, the reading being considered the true temperature of the evaporating layer.

In investigating convection in a shallow layer of silicone oil under an air surface, Koschmieder[6] used copper-constantan thermocouples for temperature measurements. The dependence of surface tension on temperature was found to be -0.058 dyne/cm·°C ($\pm 5\%$).

Kuehn and Weaver[7] described an apparatus for measuring and plotting the temperature distribution in the vicinity of an evaporating (diffusion-controlled evaporation) liquid surface, claiming a precision of ± 0.01°C. The apparatus consisted of a small thermistor forming one arm of a Wheatstone bridge, a device to lower the thermistor through the liquid surface, a coupled voltage divider to monitor the position of the apparatus, and recording instrumentation. In air at 21°C and 65% RH and with a water body temperature (measured with a mercury thermometer) of about 18°C, the depression of the temperature at the water surface with respect to the bulk water was found to be 0.83 ± 0.04°C compared with the corresponding theoretical value (presumably calculated by the method attributable to Schirmer[8]) of 0.74 ± 0.04°C.

Ewing and McAlister[9] investigated the thermal boundary layer of the ocean by making simultaneous radiometer and thermistor measurements. The radiometer temperature determination was more than 0.6°C cooler than the thermistor determination, under conditions of wind and humidity which the authors stated were "not conducive to vigorous evaporation." Laboratory experiments were performed also in a controlled environment in which the chief features of the ocean were modeled, with the general result that the departure of the radiation temperature measured with the radiometer from the bulk water temperature increased with the flux of heat through the radiation layer, a layer of water extending from the surface down about 0.1 mm (0.004 in.). It was concluded that the radiometer measurements demonstrated the existence of a very thin cool layer on the evaporating ocean as much as 0.6° cooler than the conventional "surface temperature" measurements.

Jarvis and Kagarise[10] investigated the temperature of a water surface across which a stream of dry gas, of less than 1% relative humidity, was passed, using a commercial infrared radiometer and thermistor probes. The experiments were performed in a room maintained at 20.0 ± 0.4°C and 50 ± 5% relative humidity. The authors concluded that both methods were satisfactory for following changes in surface temperature for evaporation studies. The radiometer measured a much thinner layer of water (0.1 mm, 0.004 in.) than did the thermistor (2 or 3 mm, 0.08 or 0.12 in.), responded more rapidly, and could be operated away from the surface. The thermistor had a higher sensitivity, 0.01°C compared to 0.1°C or greater for the radiometer. Within the limit of sensitivity of the radiometer the presence of a monomolecular layer film (oleic acid) did not significantly alter the emissivity of the water surface, indicating that radiometer measurements of surface temperature changes in the presence of such a film would be measurements of changes in the temperature of the water surface.

Eagleton et al.[11] used an infrared radiometer to measure water surface temperature, noting that a radiometer is sensitive to the average temperature of the top 0.1 mm (0.004 in.) or less of the liquid, whereas a thermistor immersed in the liquid in the presence of thermal gradients might provide a poor measurement of the surface temperature. With the radiometer and a 0.25-mm (0.01-in.) diameter thermistor, simultaneous measurements of surface temperature during evaporation gave values for the radiometer approximately 0.2 to 0.5°C lower than those for the thermistor. It was stated that the interpretation of the thermistor measurements was complicated by the presence of the thermistor in the liquid and by the meniscus effect. In later work,[12] a 0.125-mm (0.0049-in.) diameter thermistor was placed at the surface, about one third of it in the liquid and about two thirds above the flat liquid surface. The top of the thermistor was kept wet by surface tension effects. The water evaporated from a thin layer about 2 mm (0.08 in.) thick and 1.4 cm (0.55 in.) in diameter.

Grossman et al.[13] reported aircraft-borne infrared radiometer measurements of water surface temperatures of Lake Hefner, Oklahoma, 0.4 to 0.9°C cooler than the water about 1 cm (0.4 in.) below the surface. Under a monomolecular film (hexadecanol-octadecanol) used to reduce evaporation, the water surface temperature was reported to be 0.3°C warmer than the surrounding free surface. The absolute error of the radiometer was reported to be ±0.5°C and the relative error was reported to be ±0.1°C.

Lorenz[14] reviewed the measurement of the surface temperature using the infrared radiometer. The emphasis was on airborne measurements. Possible errors in the method were discussed and included the following sources: (1) targets which are not blackbody radiators; (2) uncertainties in the spectral distribution of the emissivity or reflectivity of the target; (3) absorption and emission in the air layer between the target and the detector; (4) inclination of the radiometer from the normal to the target surface; and (5) the nonuniformity of the target surface. Lorenz gave an example of airborne measurements of the surface temperature of a lake and of land. The corrected measurements for the lake were within less than 1°C of bucket measurements of water temperature made at the same time from a boat on the lake.

In the experiments of Johnstone and Smith,[15] a Mach-Zender interferometer[16] was used to measure the nonsteady change in temperature. The method is based on the temperature of the liquid over a short range of temperature. The optical path length is proportional to the index of refraction; therefore, the changes in optical path length measured with the interferometer were proportional to the changes in temperature. The accuracy of the temperature measurement, equivalent to $^1/_{20}$ of a fringe at light wavelength of 5461 Å (2.150×10^{-5} in.) and a path length of 4.75 in., was claimed to be about 0.003°C. The accuracy of the measurement of the liquid surface temperature using the interferometer was claimed to be about 0.005°C.

Adams and Meier[17] used a Mach-Zender interferometer with a helium-neon gas laser light source to investigate density variations in distilled water near the water/air interface. Two conditions were investigated: (1) evaporation from an undisturbed interface and (2) evaporation from an interface disturbed by the presence of a submerged, heated copper tube. Variations in the interfacial temperature, changes in the thickness of the interfacial thermal layer, and temperature gradient at the interface were investigated. Adams and Meier concluded that small momentum disturbances at the underneath side of the thermal layer were responsible for triggering a reaction in the unstable layer, which altered the frequency and magnitude of the small temperature changes that would normally occur in the absence of a submerged, heated surface.

Alty[18] determined surface temperature from the measurement of the surface tension of the evaporating surface, using the argument that the surface temperature must be controlled by the temperature of the outermost layers of the liquid and that the temperature derived from the determination of surface tension should be a very close approximation to the surface temperature required for evaporation work. To measure surface tension, he used the drop-weight method described by Harkins and Brown,[19] in which drops of air-free distilled water were allowed to form slowly in an evacuated space on a specially prepared tip and after falling under gravity were collected. The surface tension was derived from the weight of the drop by the formula of Harkins and Brown relating surface tension to the drop mass, the acceleration due to gravity, the volume of the drop, and the radius of the tip. Alty deduced the surface temperature from the surface tension using the data of Young and Harkins.[20]

Archer and La Mer[21] used thermometers to determine the surface temperature from an estimation of the temperature gradient in the water.

Jarvis et al.[22] used a thermistor placed beneath the water surface to record the average temperature of a layer of water the thickness of the thermistor bead. The thermistor was moved closer and closer to the surface to determine the approximate temperature profile of the water. By extrapolation of the curve for the clean water surface to zero average thermistor depth, an approximate value for the surface temperature of the water was obtained.

Baranev[4] probed an evaporating surface with a fine thermocouple. As the liquid evaporated, it eventually overtook the thermocouple, the temperature indication of which had been falling steadily. As the thermocouple and the "surface" coincided, the indication remained steady as long as the thermocouple could maintain a meniscus. This indication was considered to be the true temperature of the evaporating layer.

Maa[23] used jet tensimeter data to calculate the liquid surface temperature.

Paulsen and Parker[24] monitored the temperature at an effective depth of 20 μm (7.9×10^{-4} in.), taken to be surface temperature, using a radiation thermometer operating in a wavelength band of 8 to 15 μm.

Davis et al.[25] used the data of Maa to predict the interfacial temperature in a rigorous manner, without recourse to iterative procedures.

O'Brien et al.[26] used a Fabry-Perot interferometer to measure the temperature profile of surface water layers, and to determine the drop in surface temperature with evaporation.

Katsaros and Garrett[27] read interface temperature with an infrared radiation thermometer to 0.1°C accuracy. The fluctuating temperature in the water was measured with a resistance film probe.

Salhotra et al.[28] used thermometers to measure the temperature at the water surface; at various depths within evaporation pans.

Chattree and Sengupta[29] used an infrared radiation thermometer to measure the surface temperature of water.

Miller and Millis[30] used thermal infrared satellite imagery to determine water surface temperature.

Hickman and Kayser,[31] addressing the difficulty in determining the temperature precisely at the interface of a liquid with a gas, usually air, described an indirect method of "great precision that can be applied to volatile liquids if the operation can be conducted in a closed system with exclusion of air or other gas."

They showed that the "true temperature of the surface of a clean volatile liquid, for example, water, in contact with its saturated vapor can be inferred with a certainty of 0.002°C from barometeric readings under ambient conditions even though heat energy may be inserted or removed equivalent of 0.1 W/cm^2 of surface." The system and method permitted determination of surface tension and other physical characteristics.

TEMPERATURE DIFFERENCE BETWEEN WATER SURFACE AND BULK

Paulson and Parker[24] examined the temperature difference between a water surface and 10 cm (3.9 in.) depth in the laboratory as function of surface stress and heat transfer just below the gas-liquid interface.

The sea surface temperature is usually cooler than the water immediately below, because the upward heat flux in the water just below the interface is most often positive. Saunders[32] hypothesized the existence of a layer immediately below the interface in which the vertical transfer of heat and momentum is primarily by molecular processes. The temperature difference, ΔT, between the surface and a lower well-mixed region of nearly constant temperature is a function of a dimensionless constant determined from observations, the upward heat flux, the kinematic viscosity of water, the thermal conductivity of water, the shearing stress in the air above the water, and the density of water. These last three quantities are in the denominator of the expression for ΔT. McAlister and McLeish[33] suggested that the thickness of the layer is of the order of 1 mm (0.04 in.).

The purpose of the Paulson and Parker paper was to describe a laboratory experiment designed to evaluate the dimensionless constant, λ, to test the prediction that λ ought to be independent of variations in upward heat flux and shearing stress, and to compare the results with other laboratory and field measurements.

The experimental apparatus consisted of an insulated water-filled container over which air was blown at a constant velocity. The bulk water temperature was measured at a depth of about 10 cm (3.9 in.) using a mercury thermometer. The temperature at an effective depth of about 20 μm (7.9×10^{-4} in.), taken as the surface temperature, was monitored by a radiation thermometer. The average wind speed over the cross section of the tunnel was measured using a propeller anemometer. The humidity and temperature of the air were measured by an Assman psychrometer located at the entrance to the tunnel. Wave generation in the liquid surface was negligible, and the fetches were short.

Within the uncertainty of the observations, λ appeared to be independent of variations in stress and heat flux as predicted by Saunders[32] and by Hasse.[34] The average value of λ (equal to 15) was larger than the value of 5 to 10 suggested by Saunders on the basis of limited field data.

TEMPERATURE GRADIENTS

A 20-junction chromel-P-constantan thermopile of 0.003-in. diameter junctions was used by Spangenberg and Rowland[35] to measure temperature differences in water to the nearest 0.01°C. One set of the junctions was movable throughout the container with a screw-controlled traversing assembly; the other set was fixed in a massive holder near the bottom of the container.

To measure air temperature and liquid temperature, mercurial thermometers calibrated to 0.1°C were used. Since the presence of the thermopile affected the results, i.e., the convective pattern, some of the measurements were taken as a function of time. The thermopile was placed at fixed distance below the liquid surface; and the water temperature difference was measured as a function of time elapsed after the container was uncovered, requiring a separate cycling of evaporative cooling for each depth. Transverse temperature distributions within a plunging sheet (Spangenberg and Rowland's description of the appearance, in the schlieren image, of liquid "plunging" from the cooled surface layer toward the bottom of the container) were measured by fixing the thermopile in position below the surface and taking temperature readings as the sheet drifted past the thermopile.

Shiba and Ueda[36] measured the distribution of temperature near the surface of a glass plate covered with wet blotting paper using a copper-constantan thermocouple mounted on a movable device with a vernier.

Leontiev and Kirdyashkin[37] used two nichrome-constantan thermocouples 0.06 mm (0.002 in.) in diameter to measure the temperature of a glass surface. The junctions were set flush with the glass surface and were glued to the glass with epoxy resin, and then the glass was baked at 120°C. The temperature of a copper heat-transfer surface was measured with nichrome-constantan thermocouples, 0.2 mm (0.008 in.) in diameter, sealed into the copper plate. For the measurement of temperature over the volume of a separate convective cell, "combs" of nichrome-constantan thermocouples 0.06 mm (0.002 in.) in diameter mounted on a stainless steel plate which permitted horizontal movement were used. There were five thermocouples in the vertical "comb" and nine thermocouples in the horizontal comb. The distances between the thermocouples in the vertical comb and the coordinates in the thermocouple junctions in the flow pattern were determined photographically. The horizontal comb was used only for measuring temperatures in the middle section of the water layer. It was noted that the presence of the thermocouples did not cause noticeable disturbances in the convective cell.

Navon and Fenn,[38] in measuring temperature profiles, used a vertical array of seven thermocouples made of 0.005-in. diameter copper-constantan wires stretched horizontally between the two legs of a U-shaped Teflon mount. The measuring junctions were located at the midpoints of the mounted wires. Vertical and horizontal temperature distributions could be measured by moving the mount with a lead-screw device. The reference junction was located at a position in the water where the water temperature was unaffected by the evaporation. It was claimed that temperature variations of 0.02°C were easily detected.

For determining the temperature profile in evaporating propyl alcohol at the onset of convection (as detected by a schlieren image), Vidal and Acrivos[39] used a thermocouple junction mounted at fixed depths of the liquid layer, of 0.5-mm (0.02-in.) intervals, and the evaporation procedure repeated for each depth.

Yen and Galea,[40] in investigating the onset of convection in a water layer formed continuously by melting ice by determining the time at which the temperature profile of the water layer began to deviate from linearity, used bead thermistors with a bead diameter of 0.08 cm (0.03 in.) mounted in bakelite tubing, fixed in place by silastic with the beads just outside the tubing opening. To determine the temperature profile, the thermistors were slowly lowered through the liquid layer in 0.25-cm (0.10-in.) increments until they reached the water-ice interface.

Deardorff et al.,[41] to measure mean temperature at various heights in a cylindrical tank of water, used a long platinum alloy resistance wire of 0.005 cm (0.002 in.) diameter strung back and forth across a horizontal brass ring support (inside diameter 52 cm, 20 in.) in such a manner than the ratio of the wire length in a given annulus to the area of that annulus was nearly constant.

Jarvis[42] measured temperature at three depths in a water tank using thermistors mounted in waterproof glass probes approximately 1 cm (0.4 in.) long with a maximum diameter of 0.2 cm (0.08 in.). The three thermistors were mounted, one at a depth of 15 cm (5.9 in.), one at a depth of 3 cm (1.2 in.), and one near the surface. The average depth of water over the surface thermistor was about 2 mm (0.08 in.) during most of the studies, but the depth could be varied from 0 to 3 mm (0.12 in.). Temperature changes of $\pm 0.01°C$ could be detected.

Two thermocouples (presumably copper-constantan) were located in the same plane as the resistance wire, one in the center of the tank and the other at a radius of 19.3 cm (7.6 in.). The ring support could be moved vertically at rates of 1.0 cm/sec upward and 1.1 cm/sec downward. The depth of the water was about 35 cm (14 in.). The resistance wire was cycled repeatedly to obtain successive mean temperature profiles during the experiment. From the results of a test in which the resistance wire was cycled 30 times at the same rate as during the experiment, it was concluded that the sensor and probe movement caused no significant mixing of the water.

TEMPERATURE OF WATER IN FILM-TREATED RESERVOIRS

Bartholic et al.[43] made observations concerning the temperature of water in reservoirs. The following quotations were taken from their paper.

The portion of the reservoir surface covered with a monolayer may vary from a few percent to 100%, depending on wind speed and the techniques and frequency of dispensing.

When a monomolecular film is applied to a water surface, evaporation is reduced and the energy that would otherwise have gone into evaporation goes instead to increase the temperature of the water. As the water temperature increases, energy transfer to the atmosphere from the body of water by thermal radiation (Q_r) and conduction (Q_c) increases.

As a monolayer suppresses evaporation, the energy that would have been used in evaporation goes into warming the surface. Since the surface water is now warmer relative to the lower depths of the water, some energy can be expected to be transported down to warm the lower depths. In general, then, the reservoir warms to a new temperature, which simultaneously raises the vapor pressure at the reservoir surface. This physical argument is borne out by data collected specifically to examine this affect at the 600-acre Pactola Reservoir, where, over a 98-day period, a high monolayer coverage of 73% was maintained.

(The Pactola Reservoir is in South Dakota.)

WATER VAPOR DISTRIBUTION ABOVE A WATER SURFACE

Gates et al.[44] used a microwave refractometer to make measurements of the moisture profile above a water surface. Spatial resolution was obtained by drawing air samples into the refractometer cavity through a hypodermic needle connected by plastic tubing to the cavity. The vertical positioning of the needle was accomplished by the use of a rack and pinion. Their plot of water vapor pressure against distance shows a decrease in water vapor pressure with distance which could be described as roughly exponential. It was claimed that a change in moisture concentration equivalent to a variation as small as 0.1% relative humidity at 20°C could be detected and that the response time of the instrument was of the order of 30 sec. Gates[45] used the refractometer to investigate the moisture boundary layer near plant leaves and an open water surface. He attributed an abrupt bending of the moisture profile toward the surface within the first several mm of the surface, giving the profile a sigmoid shape, to the measuring technique. He also provided schlieren photographs of boundary layers illustrating the use made of schlieren photography by Gates and Benedict[46] in connection with convection in the vicinity of plants.

Shiba and Ueda[47] used a resistance electric hygrometer (10 mm long, 4 mm wide, and 0.2 mm thick (0.39 in. long, 0.16 in. wide, and 0.008 in. thick) of otherwise unspecified type, mounted on a movable device, to measure the humidity distribution near the surface.

Jones[48,49] used a miniature barium fluoride humidity element in a preliminary investigation of the distribution of water vapor above a free-water surface. The element (1 × 2 × 0.08 cm; 0.4 × 0.8 × 0.03 in.) was mounted above a water surface in a 2000 mL Pyrex beaker with a water depth of 28 mm (1.1 in.).

The beaker rested on an adjustable support; the elevation of the element above the water surface was varied by raising or lowering the beaker. Except at the closest approaches to the surface, the recorded indication of the element fluctuated about a mean indication at fixed elevations. During the experiment, the room temperature varied between 24.3 and 25.0°C. The indicated relative humidity at the closest approach to the surface (2 mm, 0.08 in.) was 98%, decreasing to 50% at 7.5 mm (0.30 in.), and 27% at 31.5 mm (1.24 in.). The room relative humidity was 15%.

The general shape of the smooth curve drawn through the data was similar to the moisture profile of Gates et al.[44] taken by drawing air samples from above a water surface through a hypodermic needle into a microwave refractometer cavity. Presumably any fluctuations about the mean values in the Gates et al. work were not detected due to the method of sampling. This preliminary experiment indicated the possibility of using the barium fluoride element for investigating the water vapor pressure distribution above a free-water surface.

REFERENCES

1. Pasquill, F. "Evaporation from a Plane Free-Liquid Surface into a Turbulent Air Stream," *Proc. R. Soc.* A182:75 (1943).
2. Yamamoto, G. "Investigation of Evaporation from Pans," *Trans. Am. Geophys. Union* 31:349 (1950).
3. Vidal, A., and Acrivos. "Effect of Nonlinear Temperature Profiles on the Onset of Convection by Surface Tension Gradients," *Ind. Eng. Chem. Fundam.* 7:53 (1968).
4. Baranev, M. *J. Phys. Chem. (USSR)* 13:1635 (1939).
5. Hickman, K. "Reviewing the Evaporation Coefficient," *Desalination* 1:13—29 (1966).
6. Koschmieder, E. L. "On Convection Under an Air Surface," *J. Fluid Mech.* 30:9 (1967).
7. Kuehn, L. A., and R. S. Weaver. "A System for Recording Temperatures in the Vicinity of Liquid Surfaces," *J. Sci. Instrum.* 1:776 (1968).
8. Schirmer, R. *Verh. Dtsch. Ing.* 5:170 (1938).
9. Ewing, G., and E. D. McAlister. "On the Boundary Layer of the Ocean," *Science* 131:1374 (1960).

10. Jarvis, N. L., and R. E. Kagarise. "Determination of the Surface Temperature of Water During Evaporation Studies. A Comparison of the Thermistor with Infrared Radiometer Measurements," *J. Colloid Sci.* 17:501 (1962).
11. Eagleton, L. C. Saline Water Conversion Report for 1965, U.S. Department of Interior, p. 168.
12. Eagleton, L. C. Saline Water Conversion Report for 1967, U.S. Department of Interior, p. 91.
13. Grossman, R. L., B. R. Bean, and W. E. Marlatt. "Airborne Infrared Radiometer Investigation of Water Surface Temperature With and Without an Evaporation-Retarding Monomolecular Layer," *J. Geophys. Res.* 74:2471 (1969).
14. Lorenz, D. "Temperature Measurements of Natural Surface Using Infrared Radiometers," *Appl. Opt.* 7:1705 (1968).
15. Johnstone, R. K. M., and W. Smith. "Rate of Condensation or Evaporation During Short Exposures of a Quiescent Liquid," *Proceedings of the Third International Heat Transfer Conference,* Vol. 2, (New York: American Institute Chemical Engineers, 1966), 348.
16. Johnstone, R. K. M., and W. Smith. "A Design of a 6-in. Field Mach-Zehnder Interferometer," *J. Sci. Instrum.* 42:231 (1965).
17. Adams, J. A., and L. B. Meier. "Density Gradients Near a Liquid/Air Interface," U.S. Naval Research Lab. Report 6828 (1969).
18. Alty, T. "The Maximum Rate of Evaporation of Water," *Philos. Mag.* 15:82—103 (1933).
19. Harkins, W. D., and F. E. Brown. "The Determination of Surface Tension (Free Surface Energy), and the Weight of Falling Drops; The Surface Tension of Water and Benzene by the Capillary Height Method," *Am. Chem. Soc. J.* 41:499 (1919).
20. Young, T. F., and W. D. Harkins. "Surface Tension Data for Certain Pure Liquids Between 0 and 360°C and for All Types of Solutions at All Temperatures," *Int. Crit. Tables IV* 446 (1919).
21. Archer, R. J., and V. K. LaMer. "The Rate of Evaporation of Water Through Fatty Acid Monolayers," *J. Phys. Chem.* 59:200 (1955).
22. Jarvis, N. L., C. O. Timmons, and W. A. Zisman. "The Effect of Monomolecular Films on the Surface Temperature," in *Retardation of Evaporation by Monolayers: Transport Processes,* V. K. LaMer, Ed. (New York: Academic Press, 1962), pp. 41—58.
23. Maa, J. R. "Rates of Evaporation and Condensation between Pure Liquids and Their Own Vapors," *Ind. Eng. Chem. Fundam.* 9:283—287 (1970).
24. Paulson, C. A., and T. W. Parker. "Cooling of a Water Surface by Evaporation, Radiation, and Heat Transfer," *J. Geophys. Res.* 77:491—495 (1972).
25. Davis, E. J., R. Chang, and B. D. Pethica. "Interfacial Temperature and Evaporation Coefficients with Jet Tensimetry," *Ind. Eng. Chem. Fundam.* 14:27—33 (1975).
26. O'Brien, R. N., A. I. Feher, K. L. Li, and W. C. Tan. "The Effect of Monolayers on the Rate of Evaporation of H_2O and Solution O_2 in H_2O," *Can. J. Chem.* 54:2739—2744 (1976).

27. Katsaros, K. B., and W. D. Garrett. "Effects of Organic Surface Films on Evaporation and Thermal Structure of Water in Free and Forced Convection," *Int. J. Heat Mass Transfer* 25:1661—1670 (1982).
28. Salhotra, A. M., E. E. Adams, and D. R. F. Harleman. "Effect of Salinity and Ionic Composition on Evaporation: Analysis of Dead Sea Evaporation Pans," *Water Resour. Res.* 21:1336—1344 (1985).
29. Chattree, M., and S. Sengupta. "Heat Transfer and Evaporation from Heated Water Bodies," *J. Heat Transfer* 107:779—787 (1985).
30. Miller, W., and E. Millis. "Estimating Evaporation from Utah's Great Salt Lake Using Thermal Infrared Satellite Imagery," *Water Resour. Bull.* 25:541—550 (1989).
31. Hickman, K., and W. Kayser. *J. Colloid Interface Sci.* 52:578 (1975).
32. Saunders, P. M. "The Temperature at the Ocean-Air Interface," *J. Atmos. Sci.* 24:269—273 (1967).
33. McAlister, E. D., and W. McLeish. "Heat Transfer in the Top Millimeter of the Ocean," *J. Geophys. Res.* 74:3408—3414 (1969).
34. Hasse, L. "The Sea Surface Temperature Deviation and the Heat Flow at the Sea-Air Interface," *Boundary-Layer Meteorol.* 1:368—379 (1971).
35. Spangenberg, W. B., and W. R. Rowland. "Convective Circulation in Water Induced by Evaporative Cooling," *Phys. Fluids* 4:743 (1961).
36. Shiba, K., and M. Ueda. "Humidity Distribution and Rate of Evaporation of Water," in *Humidity and Moisture, Vol. 2,* A. Wexler, Ed. in Chief, E. J. Amdur, Ed. (New York: Reinhold Publ. Co., 1965), p. 349.
37. Leontiev, A. I., and A. G. Kirdyashkin. "Experimental Study of Flow Patterns and Temperature Fields in Horizontal Free Convection Liquid Layers," *Int. J. Heat Mass Transfer* 11:1461 (1966).
38. Navon, U., and J. B. Fenn. "Interfacial Mass and Heat Transfer During Evaporation: 1. An Experimental Technique and Some Results with a Clean Water Surface," *AIChE J.* 17:131—136 (1971).
39. Vidal, A., and A. Acrivos. "Effect of Nonlinear Temperature Profiles on the Onset of Convection by Surface Tension Gradients," *Ind. Eng. Chem. Fundam.* 7:53 (1968).
40. Yen, Y.-C., and F. Galea. "Onset of Convection in a Water Layer Formed Continuously by Melting Ice," *Phys. Fluids* 12:509 (1969).
41. Deardorff, J. W., G. E. Willis, and D. K. Lilly. "Laboratory Investigation of Non-Steady Penetrative Convection," *J. Fluid Mech.* 35:7 (1969).
42. Jarvis, N. L. "The Effect of Monomolecular Films on Surface Temperature and Convective Motion at the Water/Air Interface," *J. Colloid Sci.* 17:512 (1962).
43. Bertholic, J. F., J. R. Runkles, and E. B. Stenmark. "Effects of a Monolayer on Reservoir Temperature and Evaporation," *Water Resour. Res.* 3:173—179 (1967).
44. Gates, D. M., M. J. Vetter, and M. C. Thompson, Jr. "Measurement of Moisture Boundary Layers and Leaf Transpiration with a Microwave refractometer," *Nature* 197:1070 (1963).

45. Gates, D. M. "The Measurement of Water Vapor Boundary Layers in Biological Systems with a Radio Refractometer," in *Humidity and Moisture, Vol. 2*, A. Wexler, Ed. in Chief, E. J. Amdur, Ed. (New York: Reinhold Publ. Co., 1965), p. 33.
46. Gates, D. M., and G. M. Benedict. "Convection Phenomena from Plants in Still Air," *Am. J. Bot.* (July 1963).
47. Shiba, K., and M. Ueda. "Humidity Distribution and Rate of Evaporation of Water," in *Humidity and Moisture, Vol. 2*, A. Wexler, Ed. in Chief, E. J. Amdur, Ed. (New York: Reinhold Publ. Co., 1965), p. 349.
48. Jones, F. E. "Barium Fluoride Film Humidity Element Calibration Analysis, Applications and Other Developments," National Bureau of Standards (U.S.) Report 10058, 1969.
49. Jones, F. E., and A. Wexler. "A Barium Fluoride Film Hygrometer Element," *J. Geophys. Res.* 65:2087 (1960).

CHAPTER 7

Wind Tunnel Investigations of Evaporation

INTRODUCTION

Evaporation in the atmosphere is strongly dependent on the wind, therefore, a complete laboratory study of evaporation must include experiments in which a flow of air past the evaporating surface is carefully controlled and varied. It is immediately obvious, then, that studies in wind tunnels should be considered. In this chapter, selected wind tunnel studies will be reviewed.

SIMILARITY

Since attempts have been made to model the atmospheric boundary layer in the wind tunnel through the concept of similitude, it is appropriate here to present a digression on similitude and dimensional groups.

The definitions by Sutton[1] are pertinent here: "Geometric similitude between two systems means that one may be made to coincide with the other by a suitable change in the unit of length; dynamic similitude means that, by suitable changes in the fundamental units (those of mass, length, and time), the equations of motion and the boundary conditions of one system can be transformed into those of the other." Similitude, or similarity, is expressed by various dimensionless groups, or dimensionless numbers, arrived at through dimensional analysis. In what follows, dimensionless groups of interest in wind tunnel modeling of the atmospheric boundary layer for the study of the evaporation of water will be enumerated and defined.

Reynolds number, Re

$$\begin{aligned} \text{Re} \equiv \text{inertial forces/viscous forces} &= \rho V^2/(\mu V/L) \\ &= VL/(\mu/\rho) = VL/\nu \end{aligned}$$

where V = a characteristic velocity (free-stream velocity, for example)
L = a characteristic length
ν = the kinematic viscosity of the fluid (the ratio of the viscosity, μ, to the density, ρ)

A necessary condition for dynamic similitude where geometric similitude has been established is that the two systems have the same Reynolds number and are subject to the same boundary conditions.[1]

Prandtl number, Pr

$$Pr \equiv \nu/k$$

where k = the thermal diffusivity

The Prandtl number is the ratio of the momentum diffusivity to the thermal diffusivity and indicates the relative ease of transport of momentum and heat.

Schmidt number, Sc

$$Sc \equiv \nu/D$$

where D = the diffusivity of matter

The Schmidt number indicates the relative ease of transport of momentum and matter.

Péclet number, Pé

$$Pé \equiv VL/k;\ Pé = Pr \cdot Re$$

The Péclet number indicates the relative effectiveness of convective and molecular motion.

Nusselt number, Nu

$$Nu \equiv \ell\Theta_h/k'$$

where Θ_h = the heat transfer coefficient
ℓ = the thickness of the liquid layer
k' = the thermal conductivity

The Nusselt number expresses the ratio of the total heat transfer to that due to conduction alone.

Froude number, Fr

$$Fr \equiv V/\sqrt{\delta g}$$

where δ = the boundary layer thickness
g = the acceleration due to gravity

The Froude number squared is the ratio of the inertia force on a fluid element to the force on a fluid element due to gravity. The Froude number indicates the influence of gravity on the form of the free surface.

Grashof number, Gr

$$Gr \equiv g\beta \, \Delta T \, L^3/\nu^2 T$$

where β = the coefficient of cubical expansion of the fluid
ΔT = a characteristic temperature difference (T, absolute)

The Grashof number indicates the influence of gravity on the initiation of natural convection in the fluid.

Rayleigh number, Ra

$$Ra \equiv g\beta\zeta\ell^4/k\nu \quad Ra = Gr \cdot Pr$$

where ζ = the temperature gradient in the fluid
ℓ = the depth of the fluid

The Rayleigh number serves as a criterion for the onset of convective currents in the liquid.

Marangoni number, M

$$M \equiv \psi\zeta\nu^2/\mu k$$

where ψ = the rate of change of surface tension with temperature elevation at the surface temperature

The Marangoni number serves as a criterion for the onset of surface tension-driven convection in the fluid.

This list of dimensionless numbers does not, of course, exhaust all the possibilities, but it does include the most important groups.

WIND TUNNEL EXPERIMENTATION

Hinchley and Himus[2] used a wind tunnel to study the evaporation of water from copper pans, 3 in. deep, of four sizes with heating coils on silica tubes set along the lengths of the pans about $^1/_4$ in. from the bottom. The water temperature was varied in the range 20 to 70°C, and the air velocity was varied from 0.9 to about 5.36 m/sec. The amount of water evaporated was determined by having the pans supported on one pan of a balance. The rate of evaporation was found to be given by:

$$W = (0.031 + 0.0135\ v)(p_e - p_d) \quad (1)$$

where W = the rate of evaporation (kg/m^2·hr corrected to a barometric pressure of 760 mm of mercury)
v = the air velocity (m/sec)
p_e = the water vapor pressure of the water in the pan (mm of mercury)
p_d = the vapor pressure in the air (mm of mercury) determined by a Hinchley hygrometer

The existence of a very small edge effect (i.e., the edges of the pan interfered with the air flow) was noted. Ripples (always present at the higher velocities) on the water surface and vibration of the apparatus produced increased evaporation. Himus'[3] reanalysis of the data of Kinchley and Himus yielded the formula:

$$W = 0.02\ (p_e - p_d)^{1.2} + 0.0446\ d^{0.77}\ (p_e - p_d)^{1/1.2} \quad (2)$$

where d (not subscript) corresponds to v of the previous equation

Millar[4] performed experiments on the evaporation from free water surfaces in two wind tunnels. In the larger of the two tunnels, a copper evaporating pan rested in a larger pan of water. The edge of the evaporating pan projected about 2 mm (0.08 in.) and occasionally as much as 6 mm (0.24 in.) above the surface of the water in the larger pan. The water in the evaporating pan was continuously stirred mechanically and the rate of evaporation was determined from volumetric measurement of the water loss. In the smaller tunnel, used to perform experiments for a wider range of velocities and a smaller evaporating surface, an evaporating pan was sunk with the rim flush with the bottom of the tunnel and with the water surface within 2 mm (0.08 in.) of the top of the pan. The water temperature was averaged without being thermostatically controlled. The evaporation rate was determined by weighing. Room air was drawn through both tunnels. The theory of turbulent mass exchange was used to develop equations for evaporation from both finite and infinite free water surfaces; the formulas obtained were verified by the observational data. Millar attributed the factors involving vapor pressure in Equation 2 to the large differences in temperature between air and water in the Hinchley and Himus data.

Powell and Griffiths[5] performed wind tunnel investigations of the evaporation of water from a saturated linen sheet as a horizontal plane surface and as a vertical cylinder under the influence of steady winds in the range of velocity 0.6 to 2.6 m/sec. The rate of total evaporation for plane surfaces was found to be proportional to the 0.77 power of the length of the surface in the direction of air flow.

Pasquill[6] investigated the evaporation from plane free-liquid surfaces into a tangential airstream in a wind tunnel, demonstrated the importance of the type of boundary-layer flow, and tested the experimentally determined rates of evaporation into a turbulent boundary layer against a hydrodynamical theory of Sutton.[7] To avoid spillage from a pan and the problems associated with attempting to avoid spillage, Pasquill simulated a free-water surface using filter paper initially flooded with liquid and justified the simulation on the observation of Powell[8] that the vapor pressure at the surface of moistened filter paper and similar materials does not differ appreciably from that at the surface of free water. The loss of water due to evaporation was measured by an indicating balance set up near the wind tunnel.

In a preliminary experiment on the evaporation of bromobenzene in the wind tunnel, Pasquill found a marked rise in the rate of evaporation at a wind speed which appeared to depend on the position of the evaporating strip, and identified the effect with the transition from laminar to turbulent flow. He later demonstrated the existence of critical conditions of flow by photographic observations of smoke. He concluded that it was necessary to direct attention to the aerodynamics of the evaporation experiment and referred to the work of Skljarenko and Baranajew[9] who found sudden changes in evaporation rates for certain fluid with transition from laminar to turbulent flow in the airstream.

Pasquill's complete investigation included the determination of the rate of evaporation of bromobenzene as a function of wind speed (in the range 1 to 9 m/sec) for circular, squared, and rectangular surfaces, followed by determinations of rates of evaporation of bromobenzene, aniline, methyl salicylate, and water from the circular area at constant wind speed. The wet- and dry-bulb temperatures for the water experiments were observed using an Assmann psychrometer mounted upstream of the evaporating surface. Pasquill compared the results for the determination of the rate of evaporation of bromobenzene as a function of wind speed with calculations using the theory of Sutton[7] and found the agreement to be excellent.

Yamamoto[10] used a wind tunnel to investigate evaporation from a water surface covered by blotting paper and compared empirical equations relating the rate of evaporation and wind velocity to theoretical results. The wind tunnel was of the Eiffel type with a diameter of 25 cm (9.8 in.) at the end of the converging collector, fitted with grids of various mesh sizes covered with different kinds of cloth to obtain wind velocities in the range 0.5 to 35 m/sec. The blotting paper, used to avoid waves and spray, was in contact with the water surface in four nickel-coated brass vessels, of dimension 10 cm (3.9 in.) across the wind and 1 cm (0.4 in.) deep, with the leading edge and rear edge streamlined. The wetted surfaces were of 5, 10, 15, and 20 cm (2, 3.9, 5.9, and 7.9 in.) along the wind for the four vessels. An Assmann psychrometer inserted in the working section of the tunnel was used to measure the temperature and the water vapor concentration of the air. An experimentally determined mean value of 1.08 for the ratio of the rate of evaporation from the blotting paper to the rate of evaporation from a free-water surface was used to reduce the data to a free-water surface. For evaporation into still air, Yamamoto found the empirical relationship, of the form of Dalton's law, between the rate of evaporation in still air by natural convection and (C_0 − C_1):

$$E_0 = 0.298\ (C_o - C_1) \tag{3}$$

where E_0 = the rate of evaporation
C_0 = the water vapor concentration on the water surface
C_1 = the water vapor concentration (absolute humidity) in the air

The effects of wind velocity and the size of the vessel were examined at the same time in the form of the Reynolds number, Re:

$$Re = u_1 x/\nu$$

where u_1 = the wind velocity
x = the distance measured along the evaporating surface
ν = the kinematic viscosity of the air

The results indicated laminar flow in the range $\log_{10}$ Re less than 4.75; beyond this value of Re the results seemed to indicate turbulent flow. The inferences regarding the type of flow were drawn from a comparison of the experimental results with theoretical-empirical lines representing a function of evaporation rate vs Reynolds number. Another experiment was performed in a wind tunnel of the Gottingen type, which had an elliptic cross section at the end of the converging collector. The vessels used had experimental surfaces of length 40, 100 to 120 cm (16, 39 to 47 in.) in the direction of the wind, 40 cm (16 in.) wide, and 5 cm (2 in.) deep, and were streamlined at both ends. The range of wind speed was 18 to 35 m/sec. The rate of evaporation was determined using a balance mounted on the tunnel. In the range $\log_{10}$ Re greater than 4.75 the results were almost in agreement with a theoretical line for turbulent flow. A very marked region of transition from laminar to turbulent flow was not found, perhaps due to the location of the experimental surface. Yamamoto derived an expression for the rate of evaporation from a free-water surface:

$$E_0 = [0.298 + 0.042\ u_1(Re)^{-0.2}]\ (c_0 - c_1) \tag{4}$$

from his theory and the experiment of Millar.[4]

Cermak and Lin[11] studied evaporation from a smooth plane boundary in a wind tunnel, in which forced convection by fluid flow parallel to the boundary was the principal cause of vapor transport and compared the experimental data with results obtained using the mass transport theory of Sutton. Smooth porcelain about $^1/_8$ in. thick was used as the evaporating boundary. The porcelain plates, referred to as "main plates," covered aluminum evaporating pans containing distilled water with a depth of about $^1/_4$ in. Strips of "buffer plates" were placed along both sides of the "main plates" to eliminate lateral diffusion. Since the width of the "main plates" was relatively small, evaporation from the "main plates" with the "buffer plates" wet was considered to be a two-dimensional case. Temperature profiles and humidity profiles were taken at the rear of the wetted boundary. Psychrometers were used for humidity measurements. Experiments were performed for boundary layer flow either in transition or in the turbulent regime. Dimensionless numbers of interest were the Reynolds and the Richardson numbers.

Two forms of the Reynolds number were defined, $Re_{x'} \equiv U_0x'/\nu$ and $Re'_* \equiv U'_*x/\nu_e$, where U_0 is the ambient velocity of the mean flow, U'_* is the mean apparent shear velocity at the downstream end of the surface under consideration, x' is the distance downstream from the leading edge of the boundary, x is the distance downstream from the beginning of the evaporation boundary, ν is the kinematic viscosity of the air (molecular diffusivity coefficient for momentum transfer in air) and ν_e is the molecular diffusivity coefficient for water vapor into air. In the experimental study, Re'_* was limited to about 10^5.

Two forms of the Richardson number (a form of the Richardson number was used only as a parameter indicating the role of gravitational forces when a vertical gradient of density exists):

$$Ri \equiv (T_0 - T_m)gx/T_oU_o^2$$

and

$$Ri_* \equiv (T_0 - T_m)gx/T_0(U'_*)^2$$

where T_0 = the mean temperature of the ambient air
T_m = the mean temperature of the evaporation surface over a length x
g = the acceleration due to gravity

Plate temperature was measured using a copper-constantan thermocouple, made from No. 30 wire, located at the center of each main plate at plate surface elevation. Experimental values of the evaporation parameter,

$N \equiv Ex/(\Delta c\ \nu_e)$, were determined from evaporation runs. E is the evaporation rate (kg/m^2·sec), $\Delta c = C_i - C_0$, C_i is the saturation water vapor concentration at temperature T_m, and C_0 is the water vapor concentration of the ambient air.

According to Cermak and Lin, the theories of two-dimensional evaporation by turbulent exchange are based on the assumptions that: (1) the exchange coefficient for momentum transport is the same as that for vapor transport and (2) the pressure gradient along the evaporation boundary is negligible, which implies that the shear stress is constant in a profile. In this study, evaporation was considered to be essentially a phenomenon of turbulent exchange. The modification of Sutton's theory suggested by Pasquill consisted of using ν_e rather than ν. The following observations by the authors are of general interest: (1) in the atmosphere, the momentum boundary layer may be much thicker than the water vapor boundary layer; (2) in wind tunnel studies, the water vapor boundary layer may be thicker than the momentum boundary layer; and (3) in a wind tunnel, large-scale mixing like that in the atmosphere does not exist. Among the conclusions for the study were (1) "The evaporation parameter N may be considered as uniquely determined by the Reynolds number, Re'_*, regardless of the regime of the boundary layer flow" and (2) "Sutton's theory of turbulent exchange as modified by Pasquill proves to be a sound basis for theoretical study of mass and heat transport by forced convection."

Arya and Plate[12] investigated similarity between the stably stratified atmospheric boundary layer and a boundary layer developing over a cold plate in a wind tunnel. Controlled laboratory measurements were used to check some fine points of the similarity theory of Monin and Obukhov[13] and to bring out the precise form of the universal functions of z/L, where z is height and L is the stability length defined by:

$$L \equiv -u_*^3 g H_0 / T_a k_v \rho_0 C_p \tag{5}$$

where
- u_* = the shear velocity
- k_v = von Karman's constant
- T_a = the absolute temperature of the layer under consideration
- H_0 = the wall heat flux
- ρ_0 = the density
- C_p = the specific heat of air at constant pressure

The basic assumptions of the similarity theory of Monin and Obukhov as stated by Arya and Plate are that the flow is plane homogeneous and that vertical fluxes are constant. The first of these conditions was realized, in the sense that velocity and temperature profiles did not change noticeably with distance in the direction of flow, in the experiments. Data from a layer of z/δ between and including 0.01 and 0.15 were considered although fluxes (momentum and heat) varied up to 50% in this layer. At a distance of 24 m (79 ft) from the leading edge of the plate a boundary thickness, δ, of about 70 cm (28 in.) was obtained. The authors concluded that the surface layer of the stably stratified atmosphere could be modeled in the wind tunnel with the following requirements: (1) that the modeling be done in the lower 15 to 20% of the boundary layer and (2) that the ratio of the height, z, to the Monin and Obukhov length, L, be the same in field and model. They stated, "The experimental results are generally in excellent agreement with field data if scaled according to the similarity theory of Monin and Obukhov, thus providing both a proof for the validity of the theory and a justification for modeling of the stably stratified atmospheric boundary in a wind tunnel."

A system of "elliptic wedge" generators and a castellated barrier to produce a simulated rough wall boundary layer was adopted by Counihan.[14] The characteristics of the simulated boundary layer were sensibly constant over the area of the working section of the wind tunnel which would be occupied by a model, and there were no significant spanwise variations of the characteristics. To produce the simulated flow, a working section length of between 4 and 5 boundary layer heights was required. Counihan concluded that the system provided a suitable method for simulating a neutral atmospheric boundary layer in a wind tunnel and that the boundary layer produced was comparable to a rough wall boundary layer.

Lawson[15] reviewed attempts to produce velocity gradients in wind tunnels, dividing the techniques used to artificially thicken a boundary layer into two classes. The first class included "1. A fine-mesh screen placed so that the plate and screen form a narrow opening into the wind. 2. Sand roughness cemented to the surface." The second class of techniques included "1. Gauzes or honeycombs; 2. Rods; 3. Flat plates; 4. Obstructions (with and without mixing)." Lawson also mentioned the attempt of Strom to use the variation of temperature with height to alter the velocity profiles in a wind tunnel.

A climatological wind tunnel, constructed for studies of the water economy of plants, was described by Weatherly and Barrs.[16] The tunnel was a closed-circuit type in which the wind speed, temperature, and humidity could be controlled independently. The wind speed could be varied between 2 and 16 ft/sec, the temperature could be varied between 10 and 40°C, and the relative humidity could be varied between 40 and 100% at temperatures near 25°C and could be maintained as low as 25% at 35°C. Heating elements, a refrigerator, and atomizers were used to establish and maintain desired values of relative humidity. A given temperature could be maintained indefinitely within ± 0.5°C or less and a given relative humidity could be maintained within ± 1% or less. Deviations in relative humidity of as much as 2% occurred during change of temperature.

Chamberlain[17] used a wind tunnel to investigate transport of gases (thorium-B vapor and water vapor) to and from grass and grasslike surfaces. The purpose of the experiments was to determine the relation between the vapor flux, the surface condition, and the mixing ratio of the vapor in the airstream just outside the immediate vicinity of the roughness elements. The evaporation of water from paper-covered artificial grass and from toweling material laid on aluminum sheet was measured at various wind speeds to determine, by comparison with the experiments with thorium-B, an estimate of the effect of molecular diffusivity on transport rate at an aerodynamically rough surface. In the water evaporation experiments, the tunnel was operated on the once-through principle with room air entering upstream of the fan. Humidity was measured by psychrometers and a dewpoint hygrometer. The rate of evaporation was determined by weighing. The ratio of the molecular diffusivity of water (taken to be 0.24 cm^2/sec) to that of thorium-B (taken to be 0.054 cm^2/sec) was 4.4. The ratio of the values of 1/B, the reciprocal sublayer Stanton number, equal to the difference between the dimensionless resistances of the boundary layer for momentum and for mass, was found to be 0.62 with some variation with surface and wind speed.

The development of a low-speed wind tunnel suitable for agricultural meteorological experimentation was discussed by Wooding.[18] The following list of applications for such a tunnel was given:[18]

- (i) *Turbulent boundary layers.* Flow structure, surface sheer stress, flow over surfaces of varying roughness, turbulence statistics of shear flow
- (ii) *Turbulent diffusion,* with application to problems of atmospheric dispersal of materials such as carbon dioxide, water vapor, spores, powders, insects; advection of these materials
- (iii) *Wind erosion,* including entrainment and saltation of soil, sand, dust, and snow; dune formation and movement; simulation of riverbed movement
- (iv) *Sheltering effects* afforded by belts of vegetation, solid and porous screens, buildings, roughened surfaces
- (v) *Evaporation and transpiration* from plants under controlled conditions of temperature, humidity and lighting; evaporation from soil and water surfaces
- (vi) *Dynamical wind effects,* such as forces on plants and artificial structures; motion of plants in wind
- (vii) *Testing and calibration of instruments* for use in micrometeorology

Selected low-speed tunnels were reviewed and design considerations were discussed. The design considerations were applied to the Pye laboratory wind tunnel. The configuration was that of an open-circuit blower tunnel using a double-inlet centrifugal fan. The design and testing of the tunnel were discussed.

An experimental investigation of the effect of artificially generated waves on the rate of evaporation of water into air was carried out in a horizontal wind tunnel.[19] Data were obtained at constant relative humidity and variable air velocity, in the absence of atomization. The recirculating wind tunnel had a rectangular cross section in the test section, 40 cm (16 in.) high, 30 cm (12 in.) wide, and a total length of 6.5 m (21 ft). The test section was 95 cm (37 in.) long and consisted of a rectangular pan of the same length, 22 cm (8.7 in.) wide and 2.5 cm (0.98 in.) deep. Air was driven by a centrifugal fan. The air was heated by a series of electrical resistances. The air velocity was measured by an accumulative anemometer. Dry- and wet-bulb temperatures were measured by 0.1°C accuracy thermometers placed in the bulk of the airstream.

Liss et al.[20] used a laboratory wind-water tunnel to determine the effect of evaporation and condensation of water molecules on the rate of air-water transfer of oxygen. The wind speed, water and air temperatures, and the humidity of the air could be controlled in the tunnel. Air velocity, temperature, and humidity were measured simultaneously by means of a multiprobe comprising a Pitot tube, a thermocouple, and an aspirating pipe leading to an hygrometer.

REFERENCES

1. Sutton, O. G. *Micrometeorology* (New York: McGraw-Hill Book Company, 1953), pp. 47—48.
2. Hinchley, J. W., and G. W. Himus. "Evaporation in Currents of Air," *Trans. Inst. Chem. Eng.* 2:57 (1924).
3. Himus, G. W. "The Evaporation of Water in Open Pans," *Trans. Inst. Chem. Eng.* 7:166 (1929).
4. Millar, F. G. "Evaporation from Free Water Surfaces," *Can. Meteorol. Mem.* 1:39 (1947).
5. Powell, R. W., and E. Griffiths. "The Evaporation of Water from Plane and Cylindrical Surfaces," *Trans. Inst. Chem. Eng.* 13:175 (1942).
6. Pasquill, F. "Evaporation from a Plane, Free-Liquid Surface into a Turbulent Air Stream," *Proc. R. Soc. (London)* A182:75 (1943).
7. Sutton, O. G. "Wind Structure and Evaporation in a Turbulent Atmosphere," *Proc. R. Soc. (London)* A146:701 (1934).
8. Powell, R. W. "Further Experiments on the Evaporation of Water from Saturated Surfaces," *Trans. Inst. Chem. Eng.* 18:36 (1942).
9. Skljarenko, S. I., and M. K. Baranajew. "Concerning the Rate of Evaporation of Liquid into Moving Air," *Acta Physichim. USSR* 4:873 (1936).
10. Yamamoto, G. "Investigation of Evaporation from Pans," *Trans. Am. Geophys. Union* 31:349 (1950).

11. Cermak, J. E., and P. N. Lin. "Vapor Transport by Forced Convection from a Smooth, Plane Boundary," CER No. 55 JECI, Report No. 9, Civil Engineering Department, Colorado A & M College, Ft. Collins, CO (1955).
12. Arya, S. P. S., and E. J. Plate. "Modeling the Stably Stratified Atmospheric Boundary Layer," *J. Atmos. Sci.* 26:656 (1969).
13. Monin, A. S., and A. M. Obukhov. "Basic Regularity in the Turbulent Mixing in the Surface Layer of the Atmosphere," *USSR Acad. Sci. Geophys. Inst.*, No. 24 (1954).
14. Counihan, J. "An Improved Method of Simulating an Atmospheric Boundary Layer in a Wind Tunnel," *Atmos. Environ.* 3:197 (1969).
15. Lawson, T. V. "Methods of Producing Velocity Profiles in Wind Tunnels," *Atmos. Environ.* 2:73 (1968).
16. Westherly, P. E., and H. D. Barrs, "A Climatological Wind Tunnel," *Nature* 183:94 (1959).
17. Chamberlain, A. C. "Transport of Gases to and from Grass-Like Surfaces," *Proc. R. Soc. (London)* A290:236 (1966).
18. Wooding, R. A. "A Low-Speed Wind Tunnel for Model Studies in Micrometeorology," Division of Plant Industry Technical Paper No. 25, C.S.I.R.O., Melbourne, Australia (1968).
19. Massaldi, H. A., J. J. Gottifredi, and J. J. Ronco. "Effect of Interfacial Waves on Mass Transfer During Evaporation of Water from a Free Surface," *Lat. Am. J. Chem. Eng. Appl. Chem.* 6:161—170 (1976).
20. Liss, P. S., P. W. Balls, F. N. Martinelli, and M. Coantic. "The Effect of Evaporation and Condensation on Gas Transfer Across an Air-Water Interface," *Oceanol. Acta* 4:129—138 (1981).

CHAPTER 8

Monomolecular Films

INTRODUCTION

Monomolecular films have been effective in laboratory studies in reducing or inhibiting evaporation from water surfaces. Attempts have been made to use such films to accomplish practical reduction of evaporation, particularly from lakes and reservoirs. Actual reductions have been much lower than the approximately 60% achieved under ideal conditions. In this chapter, the work on monomolecular films is reviewed.

REDUCTION OF EVAPORATION RATE BY MONOMOLECULAR FILMS

Rideal[1] investigated the effectiveness of unimolecular films of stearic, lauric, and oleic acid in diminishing the rate of evaporation of water. The experimental apparatus consisted of a closed inverted U-tube. One limb of the tube was filled with water and the system was then evacuated. The limb containing the water was maintained at 25 or 35°C, the other limb was maintained at 0°C. The rate of condensation in the cooled limb was considered to be the rate of evaporation.

The rate of evaporation was again determined after a minute crystal or small lens of the acid had been placed on the surface of the liquid. Equilibrium between crystal or lens and film was attained by allowing the liquid to stand at the desired temperature for several hours.

The rates of evaporation from the film-protected surface were found to be much lower than those from a clean water surface. An increase in surface concentration or film pressure retarded the rate of evaporation. Both condensed and expanded films exhibited this characteristic to a similar extent.

La Mer et al.[2] determined the rate of evaporation from a water surface from the increase in weight of a container filled with a desiccant (lithium chloride) and held at a precisely fixed distance above the water surface for a specific time. The water subphase was obtained by the double distillation of tap water from alkaline potassium permanganate solution in an all-Pyrex™ distillation assembly.

Monolayers of C_{14}, C_{16}, C_{18}, C_{20}, and C_{22} alcohols were spread on the surface of the water. Before commencing evaporation resistance measurements, several monolayers were spread and swept off to remove contaminants from the surface. The monolayer was then spread and compressed immediately to the required surface pressure. At the end of the measuring period (usually 200 to 400 sec, depending on the rate of evaporation) the desiccant box was reweighed and the resistance to evaporation of the monolayer was calculated.

From measurements with and without a monolayer, the specific resistance, r, of the monolayer was calculated from

$$r = a(w_w - w_d)(t/m_f - t/m_w) \tag{1}$$

where a = the area of water surface under the desiccant
w_w = the concentration of water vapor in equilibrium with water
w_d = the concentration of water vapor in equilibrium with desiccant
t/m_f = the reciprocal evaporation rate from surface with film
t/m_w = the reciprocal evaporation rate from surface without film

The factor $a(w_w - w_d)$ was obtained experimentally by varying the height of the desiccant box above the surface.

The final check for the purity of the alcohol samples was to measure the evaporation resistance-surface pressure isotherm. If no "kink" or point of inflection existed in the low pressure region, the sample was taken to be pure.

The importance of the chain length of the monolayer-forming molecule on evaporation, resistance, and spreading rate was emphasized. As the spreading solvent, n-hexane did not have a deleterious effect on the evaporation resistance; benzene, since it was retained to some extent in the monolayer, reduced the resistance of a monolayer at all pressures; and kerosene, although retained at low pressures, was effectively squeezed out of the monolayer at high pressures. The force involved in the formation of a hole in a close-packed monolayer is primarily that of repulsion between the chains. The Columbia University quiescent evaporation resistance-surface pressure apparatus, Evaporimeter, has been shown to be a valuable tool for specifying materials suitable for field tests in respect to many properties.

Archer and La Mer[3] studied the influence of fatty acid monolayers on the rate of evaporation of water by measuring the rate of absorption of water vapor by a solid desiccant supported above the water surface. The solid desiccant was anhydrous lithium chloride supported on a heavy waterproofed water-vapor-permeable silk cloth.

One thermometer was placed just beneath the water surface and another on the floor of the trough containing the water. The container was filled with desiccant and weighed. The lid of the container was then removed and the container was put in position above the water for a definite period of time. The container was then removed, the lid replaced, and the container reweighed. The increase in mass of the container divided by the time in position gave the rate of absorption.

> The transport of water from the liquid phase to the desiccant . . . involved evaporation from the water surface, diffusion through the separating air column and the membrane, followed by condensation or absorption at the desiccant surface.
>
> The net rate of evaporation from the water surface is the difference between the absolute rates of condensation and evaporation.

The specific resistance of the film to the evaporation of water was calculated from the difference between the reciprocals of the rates of absorption of water vapor by a solid desiccant with and without a monolayer.

> The specific resistance is defined by the product of the total resistance and the area of the film under the desiccant. The specific resistance of a monolayer is much greater than that of a clear water surface . . . always the case for the films studied.
>
> The specific resistances of monolayers of margaric acid, stearic acid, nonadecanoic acid, and arachidic acid were studied.

The specific resistance was measured as a function of surface pressure, chain length, monolayer phase, subphase composition, and surface temperature. The characteristics of a plot of specific resistance vs surface pressure were very sensitive to variations in the technique used to spread the film. Benzene and petroleum ether were used as spreading solutions. Benzene spreading solutions yielded consistently low values for specific resistance, and petroleum ether solutions yielded larger values for specific resistance.

In the liquid condensed phase, monolayer resistance was independent of surface pressure and subphase pH; its logarithm was a linear function of chain length. A theory was proposed for the source of the energy barrier. Calculation of its magnitude from heats of vaporization data agreed well with the experimental values. All of the monolayers studied decreased the rate of evaporation by a factor of about 10,000.

Shukla et al.[4] determined the efficiency of monolayers of several substances for water evaporation reduction, using a simple method that had previously been found to give a good indication of efficiency in semi-field trials.

The experimental apparatus consisted of two identical Petri dishes on two hollow brass jackets through which thermostatted water flowed continuously. Each of the dishes was filled nearly to the brim with about 80 mL (2.8 oz) and accurately weighed on an analytical balance. The evaporation retardant was spread on the surface of water in one of the dishes by careful addition of 0.2 mL (0.007 oz) of a 0.1% solution of the substance in Analar light petroleum. The temperature of the water in each Petri dish was measured using calibrated mercury-in-glass thermometers, the relative humidity of the air was determined using a wet- and dry-bulb hygrometer, and the total wind across the water surfaces was measured using an anemometer placed near the dishes. Values of percent reduction in evaporation were obtained from the loss in weight of the dishes after an exposure of about 1 hr.

The retardant monolayers investigated were of octadecyl alcohol, of compounds formed by condensing 1, 2, and 3 ethylene oxide molecules with octadecyl alcohol, and of a few typical mixtures of these. The experimental temperature range was 20 to 40°C. For all of the individual compounds except octadecyl alcohol, the percent reduction in evaporation increased more or less linearly with decreasing temperature. The reduction was not quite linear for octadecyl alcohol at the lower temperatures; this behavior was attributed to spreading difficulties. Values of evaporation reduction ranged from 15.3 to 78%. It was determined that the values probably depend on the degree of compactness of the film, i.e., on the film pressure.

The authors concluded that monolayers of products obtained by condensation of ethylene oxide with octadecyl alcohol would give better retardation of water evaporation. Compared with the original alcohol, the most efficient monolayer was that of the monoethylenoxy derivative.

La Mer and Healy[5] reviewed the subject of retardation of water evaporation by monolayers. They subtitled the article, "Spreading a monomolecular film on the surface is a tested and economical means of reducing water loss." They demonstrated conclusively that a compressed, molecularly oriented monolayer will be effective under field conditions. Australian workers then demonstrated successful methods for applying such monolayers on the surface of large reservoirs to control evaporation. La Mer and co-workers,[2] in laboratory investigations, studied the rate of evaporation in terms of the state of film compression and temperature dependence.

The transport of a molecule of water from the liquid state, through the monolayer, to the vapor state is considered to be a kinetic process involving the surmounting of a number of successive energy barriers: barriers at the water surface, the monolayer, a diffusional barrier as the water is transported through the overlying layer of air that extends to the absorbing substance used as a detector, and barriers at the absorbing interface. In the La Mer et al. work,[2] the impermeable membrane on which the absorbing desiccant rested was supported from 1 to not more than 8 mm (0.04 and 0.3 in.) above the monolayer, thus reducing the diffusion layer thickness and eliminating turbulence. The water molecules emerging from the surface of the monolayer diffused directly to the surface of the desiccant where they were condensed and weighed. The individual energy barriers act as a set of resistances in series and are therefore linearly additive.

The difference in overall resistance, as measured when there is a monolayer film and when there is no monolayer, gives the specific resistance due to the monolayer alone. The specific evaporation resistance is a property of the monolayer; it is expressed in absolute units, seconds per centimeter.

Rosano and La Mer[6] had shown that straight-chain alcohols and acids, with alkyl chains free of branching and double bonds, were the materials that formed the best condensed monolayers.

Chang et al.[7] noted the following:

> Chain length is the most important single parameter of the molecular architecture in that it controls not only the resistance to evaporation but also (i) the rate at which material spreads and forms a monolayer, (ii) 'squeezing out' of one component of a monolayer by another, (iii) 'squeezing out' or rejection of nonvolatile solvent (or solute) impurities spread with the aid of a solvent, (iv) resistance to the action of wave or wind (Healy and La Mer had presented a quantitative analysis of the effects of waves on the monolayer), (v) regeneration or sublimation of the long-chain alcohol in water, and (vi) degradation of the monolayer through the action of bacteria.

Alcohols with 16 to 18 carbon atoms were found to be the most suitable for field application.

Barnes and La Mer[8] had investigated the desirability of using mixtures of compounds with molecules of different chain lengths. At surface pressures exceeding 15 dyne/cm the alcohol-alcohol monolayers proved to be ideal.

Most impurities, including the short-chain alcohols, produced a decrease in resistance to evaporation and in surface pressure with time. Impurities which are rapidly squeezed out of the monolayer at low surface pressure had little effect on the retardation efficiency of the monolayer.

The use of benzene as a spreading solvent for the film material lessened the resistance of the monolayer to evaporation, since the benzene was retained in the monolayer. Petroleum ether (n-hexane) did not reduce the resistance of the hexadecanol monolayer at any pressure.

The rate at which a solid alcohol spread and formed a monolayer was influenced to a remarkable extent by thermal pretreatment of the alcohol sample. Studies on spreading pointed to the desirability of using well-tempered flakes of solid alcohols as monolayer materials.

At surface pressures in excess of 15 to 20 dyne/cm, and with sufficiently energetic capillary waves, the monolayer apparently becomes submerged. There is also an accompanying decrease of about 25 to 30% in the resistance to evaporation.

Since meterological data on water-storage areas are usually incomplete, and losses by seepage cannot be estimated reliably, field experiments on evaporation are difficult to assess. The natural variable that has the greatest adverse effect on evaporation control efforts is wind. Evaporation control through the use of a monolayer becomes difficult with winds stronger than 24 km/hr (15 mi/hr).

In field tests, a commercial mixture of hexadecanol-octadecanol was applied to Lake Cachuma in Southern California. The efficiency of evaporation retardation was tested by measuring the rates of evaporation from large evaporation pans, under stable atmospheric conditions. The data, an almost linear function of time, showed a reduction in the rate of evaporation of 46 to 50% when the monolayer was used.[9]

STEPS IN THE EVAPORATION PROCESS

MacRitchie[10] considered the sequence of steps in the evaporation process to be (1) adsorption, (2) vaporization, (3) diffusion, and (4) convection. He devised a simple test to determine which of the steps a monolayer affects most significantly when it retards the evaporation.

An oven with temperature controlled at 20 ± 0.2°C was used for the experimental work. An electric fan was used to generate wind at a controlled speed. The air velocity, at about 3 cm (1.2 in.) above the sample surface, was measured using a flowmeter. A lower relative humidity, about 25%, was maintained by placing large trays of calcium chloride in the oven. A higher relative humidity, about 89%, was maintained by trays of potassium chloride solution. The relative humidity in the oven was measured using a hair hygrometer. Evaporation rates were measured in aluminum trays, 9.0 cm (3.5 in.) square, the edges of which were rounded for streamlining.

For each evaporation rate measurement, the tray was filled with water to within about 1 mm (0.04 in.) from the top, placed in position on a copper block in the path of the air current, and allowed to stand for 30 min to reach steady-state conditions. It was then quickly removed from the oven and weighed on an analytical balance to 0.0001 g (0.2 micropound). Weighings were repeated at intervals of about 60 min.

Monolayers of hexadecanol were spread at their equilibrium spreading pressure by addition of a small excess of crystals to the water surface. The retardation of evaporation by the monolayer was independent of the absolute rate of evaporation, and was the same under comparable conditions of convection. Thus the hexadecanol monolayer exerted its effect by altering the hydrodynamic boundary conditions; that is, the thickness of the boundary layer was increased.

Remarkably small surface pressure gradients can cause significant changes in boundary layer properties and, consequently, in evaporation rates. The condition of no net flow of the monolayer is probably, according to MacRitchie, more closely approached by highly incompressible layers, thus explaining the effectiveness of these in reducing evaporation.

CHANGES IN THE GASEOUS DIFFUSION LAYER

MacRitchie[11] stated that, at the time his paper was written, apparently no attempts had been made to relate the effects of monolayers on evaporation with changes in the gaseous diffusion layer. The effects had usually been attributed to either an alteration in the accommodation coefficient of water or to an activation barrier to diffusion through the monolayer.

The steady-state rate of water evaporation was measured in a room maintained at 20 ± 1°C and 50 ± 3% relative humidity. The rate was measured, using an analytical balance, by the rate of weight loss from narrow tubes containing water. It was expected that conduction would be eliminated inside the tubes and the diffusion barrier could thus be varied by filling the tubes with water to different heights.

The diffusion coefficient for water vapor in air at 20°C, calculated from the experimental data, was 0.26 cm^2/sec, which agreed well with the quoted value of 0.239 cm^2/sec at 8°C.

The steady state rates of evaporation from shallow dishes were also measured with and without hexadecanol monolayers in a variety of conditions of forced convection. Forced air circulation was produced by a variable speed fan. A small excess of hexadecanol crystals was placed on the water surface in each case to maintain the monolayer at its equilibrium spreading pressure.

The hypothesis that hexadecanol acts by influencing the thickness of the stagnant layer of gas above the surface was tested by measuring the temperature gradient near the surface, using thermistor beads above the surface and a reference thermistor bead immersed in a thermostatted bath controlled to ± 0.005°C. For these experiments, large aluminum trays 30 cm (12 in.) square, were used. It was expected that the molecular diffusion gradient should correspond closely to the thermal conduction gradient. An estimate of the effective thickness of the stagnant layer was made by extrapolating the two linear portions of the temperature-distance curve.

The hexadecanol monolayer had no effect on the evaporation rate when the latter was controlled by a thick diffusion layer. Where the liquid surface was exposed to air currents, the percentage retardation of evaporation by the hexadecanol increased markedly with increasing velocity of air circulation. The close agreement between the thicknesses of the stagnant layer by two independent methods confirmed that the monolayer acted by resisting the reduction of the stagnant layer thickness by the air currents.

A comparison of a serum albumin monolayer, which is characterized by a low surface compressional modulus, with hexadecanol, which has a high surface compressional modulus, suggested that this modulus is the important factor in dissipating eddy currents near the interface, thus producing thicker diffusion layers.

CHANGES IN HEAT FLUX AND NEAR-SURFACE TEMPERATURE STRUCTURE

Katsaros and Garrett[12] studied changes in heat flux and near-surface temperature structure of water on the surface of which monomolecular films were applied.

A large convection tank, 0.75 m (29.5 in.) long, 0.50 m (19.7 in.) wide, and 0.50 m (19.7 in.) deep, was used in the study. The surface-active compounds, oleyl alcohol, 1-heptadecanol, 99%, and methyl oleate, 98%, were applied as monomolecular films. Experiments were performed at water temperatures of 20 and 28°C. In several cases the effect of a fan-driven airflow over the tank, corresponding to a wind of 1 m/sec (3.3 ft/sec), was also measured.

The temperature and humidity of the room were monitored with a ventilated Assman psychrometer. Both dry- and wet-bulb temperatures remained constant to ± 0.3°C in all the experiments. The heat flux was determined by calorimetry. A mercury-in-glass thermometer with 0.01°C marking was used to determine the temperature of the bulk fluid. The interface temperature was read with an infrared radiation thermometer to 0.1°C. The fluctuating temperature in the water was measured using a resistance film probe. The sensor was 15 microns (5.9×10^{-4} in.) in diameter and 1.2 mm (0.047 in.) long. Wind velocity was measured at the water surface position using a hand-held anemometer.

The results for the experiments showed that conditions within the bulk of the water are important, i.e., when buoyancy forces are large in the water, effects of surfactants are reduced. Even a weak airflow results in substantial changes. The heat flux becomes larger and the temperature spectrum shifts to lower frequencies. At 20°C these effects were modified by the surfactant monolayers, while at 28°C the presence of surfactants was irrelevant in both the calm and the wind-stirred condition.

Several monolayers reduced evaporation, but not so much as to eliminate the difference in temperature between the surface and the bulk water. Even a weak airflow seemed to mitigate the effects of the pure surfactants employed in this study. The implication of the study is that the internal dynamics of a water body and the external conditions in the atmosphere above are as important as the chemical properties of a surfactant in affecting the evaporation rate.

ALTERATION OF SURFACE TEMPERATURE

Jarvis[13] determined under what conditions one might expect the surface temperature of a clean water surface to be altered by the adsorption of a monomolecular film. Experiments were conducted in a glass tank 45 cm long, 25 cm wide, and 23 cm deep (18 in. long, 9.8 in. wide, and 9.1 in. deep), provided with a Lucite cover. The sides of the cover were slotted to allow movement of barriers to control compression of the monomolecular films.

Temperature measurements were made with thermistors. Three thermistors were used: one near the surface, one at 3 cm (1.2 in.), and one at 15 cm (5.9 in.). Temperature changes of ± 0.01°C could be detected and were recorded. The average depth of water over the surface thermistor was about 2 mm (0.08

in.), being easily varied from 0 to 3 mm (0.12 in.). The gas flow rate could be varied from 0 to 10 L/min and the relative humidity from approximately 0 to 95%. Room temperature and the temperature of the incoming gas were held at 20.5 ± 0.2°C, and the bulk water temperature was nearly constant at 19.20 ± 0.5°C. Movement of bulk water, as well as the surface layer, was observed during the experiments; motion of the surface and bulk water was correlated with gas flow rate and the presence of the monomolecular film.

The film material was oleic acid, cetyl alcohol, or stearic acid. Hexane or petroleum ether was used as the spreading solvent for the monolayers. Distilled water, allowed to equilibrate with the carbon dioxide in the atmosphere, was used as the substrate liquid.

Studies had shown that monomolecular films will change the temperature of the water surface, the reduced rate of evaporation giving rise to a surface warming. The monomolecular films also changed the surface tension and surface viscosity of an aqueous substrate. Jarvis's study showed that changes in these two properties can lead to changes in the water surface temperature.

At a high evaporation rate at a clean water surface, there will be a rapid cooling, generally accompanied by a convective exchange of warmer bulk water for the cooler, more dense surface water. A monolayer may change the temperature of this surface by altering the rate of this convective exchange, the change in convective flow being due to the reduction in surface tension or to the change in viscosity of the surface layer of water.

The surface temperature of a layer of water will be dependent upon many environmental factors, and can be influenced in several ways by the addition of a monomolecular film to the interface.

HEAT AND MASS TRANSFER

The approach taken by Navon and Fenn[14] in the first part of their study of the effect of monomolecular films on water evaporation was based on the unique reversal in the density-temperature relation for water which occurs at 4°C and which marks a change in the mode of heat transfer.

A vacuum sweeper probe of the type used by previous investigators was evaluated for use in the study of heat and mass transfer during evaporation. The study included both a direct experimental check of assumptions and a theoretical analysis of transfer under the effective area of the probe.

A stream of dry gas, usually nitrogen, was allowed to flow gently over a quiescent surface of water; a commercially available coulometric moisture analyzer was used to monitor the water content of the gas. A flow rate measurement permitted an instantaneous measure of the evaporation rate. Most of the heat of vaporization of the water was supplied through the water.

Since the possible effects of monomolecular films on natural convection in bulk water were to be of interest in a later study, it was desirable to determine the role of natural convection for pure water with a clean surface. To achieve this objective, a series of measurements was made for which the entire system was cooled to 0°C and then allowed to warm up. Evaporation rates and temperature profiles were followed as the temperature rose to 22°C.

The rate of evaporation rose continuously with increasing bulk and surface temperature of the water. In the vicinity of the density inversion of water at 4°C, there was a transition to bulk evaporation rate as a result of the onset of buoyancy-driven convection in the water. The Nusselt number, the ratio of total heat transferred through the water to the heat transfer due to conduction alone, rose at a bulk temperature around 4.5°C indicating the contribution due to natural convection. Thereafter, the onset of convection and the magnitude of the heat transfer were determined by the Rayleigh number, which represented the buoyancy driving force. At higher water temperatures, both the evaporation rate and the convective heat transfer rate increased as a result of the higher surface temperature.

The study showed that the measurements with the kind of vacuum sweeper used, when appropriately analyzed, can lead to the elucidation of the contribution of natural convection to the heat transfer resistance during evaporation.

INCREASE IN TEMPERATURE OF WATER IN A RESERVOIR

Bartholic et al.[15] evaluated the increase in the temperature of water in a reservoir due to the presence of a monolayer. They dealt with the assumption that the saturation vapor pressure at the water surface temperature with a film in place on the surface can be validly substituted for the corresponding value without a film in the U.S. Bureau of Reclamation simplified equation for calculating evaporation reduction.

The application of a monolayer film to the water surface of a reservoir causes a rise in water temperature, ΔT, over the corresponding temperature without a film. Using only the temperature actually measured after film application resulted in evaporation reduction calculated by the U.S. Bureau of Reclamation simplified method which was 8 to 14% too large. A method was presented for predicting ΔT which, when applied to one set of experimental data, gave a calculated value very close to the observed value of temperature change.

RESISTANCES TO EVAPORATION

Sebba and Briscoe[16] made comparative measurements of the rate of water evaporation through unimolecular films into a constant current of dry air under conditions of controlled and measured surface pressure.

The resistance offered by films to the evaporation of underlying water was a highly specific property. Stearic acid, brassidic acid, arachidic acid, cetyl alcohol, octadecyl alcohol, and n-docosanol under suitable conditions reduced evaporation to a very small fraction of that from a free water surface. Some films, e.g., of egg albumin, cholesterol, oleic acid, and elaidic acid, offered practically no resistance under any conditions.

The resistance became substantial only above a critical surface pressure which was characteristic for each substance. Hydrophilic groups in the chain appeared to reduce or eliminate the resistance. Increase of chain length increased resistance and lowered the critical pressure at which substantial resistance set in.

Sebba and Briscoe suggested that an explanation of these phenomena may require the assumption that unimolecular films contain dissolved water and are in equilibrium with a solution of the film-forming substance in the water substrate.

Cammenga and Koenemann[17] developed a method for measuring the resistances to water evaporation of monolayers to determine the resistance as a function of surface pressure in the course of a single measurement.

The Langmuir-Schaefer evaporimeter method for measuring evaporation resistances was modified to permit continuous measurements. A Langmuir trough was equipped with a Wilhelmy balance to measure surface pressure and an electronic balance was provided for the evaporimeters. Thus, the surface pressure and evaporation rate of a monolayer-covered surface of water and of a clean surface could be followed continuously and simultaneously during compression and expansion cycles for the monolayer studied. Monolayers of fatty alcohols, perfluorinated fatty alcohols, and the isomers (gluconic acid stearylamine and stearic acid glucamine) were investigated.

All measurements were made using triply distilled water. The water temperature was measured using a thermocouple. The surface pressure was measured using two Wilhelmy balances. The water evaporation rates were determined with two evaporimeters. The surface pressure and the water evaporation rate were determined simultaneously on the monolayer-covered as well as the uncovered part of the trough. After spreading, the monolayer was first compressed to a high surface pressure to eject impurities from the monolayer into the sublayer. Some minutes later the monolayer was expanded to a surface pressure of 20 mN/m; continuous measurement of the evaporation rate followed immediately.

The evaporation data for some of the fatty acid monolayers fulfilled two main predictions of theory: the logarithm of the resistance depended linearly on the surface pressure, and a series of approximately parallel lines was obtained. There was a great difference in retardation effect between the homologues of the perfluorinated fatty alcohols. No retardation effect of the stearic acid glucamine monolayer was found, the gluconic acid stearylamine monolayer produced a low but detectable evaporation resistance. Evaporation resistance increased with increasing surface pressure.

Navon and Fenn[18] investigated the reduction in evaporation from a water substrate and the accompanying change in heat transport through the water induced by monolayers of cetyl alcohol and stearic acid.

Since the experiments of Rideal in 1925,[1] the inhibiting effect that certain insoluble monolayers can have on evaporation of water has been the subject of intensive research. The available theory, suggested by Langmuir and Schaefer[19] and later developed by La Mer et al.,[2,20] explains the monolayer resistance in terms of an activation energy barrier to diffusion through the monolayer.

Evaporation from a liquid surface was considered to be governed by three rate processes, the rate of each of which was expressed in terms of a process resistance: (1) the transfer of liquid molecules across the liquid surface into the vapor phase; (2) the removal of newly formed vapor from the liquid-gas interface; and (3) the supply of heat to the surface. The evaporation rate would depend generally on the sum of these resistances. Very little effort had been made to investigate the possibility of a relation between the evaporation-inhibiting properties of monolayers and a change in heat transfer resistance of the water or a change in the diffusional resistance of the gas, caused by the presence of the monolayer. Also of particular interest is the fact that the liquid surface temperature rises when evaporation rate is suppressed.

In the Navon and Fenn work, the rate of evaporation from a given water surface area into a flowing inert gas, dry nitrogen, was measured in both the absence and presence of monolayers, using a vacuum sweeper probe. The difference between the resistances to evaporation with and without monolayer films under otherwise identical conditions yielded the monolayer resistance to evaporation.

Evaporation resistance-surface pressure isotherms were plotted for monolayers of cetyl alcohol and stearic acid. The evaporation resistance of stearic acid was found to be higher than that of cetyl alcohol. This result was attributed to the longer chain of the acid and the higher activation energy associated with the polar group.

The dependence of evaporation resistance on surface pressure at water temperatures below 15°C was reported for the first time in the Navon and Fenn work. The lower temperatures resulted in higher evaporation resistance, particularly at high compression of the monolayer. In the presence of a compressed monolayer, the natural convection in the water was reduced. Rapid drops in the Nusselt number, the ratio of the actual heat transfer to that which could be accounted for by conduction alone, indicated that the convection was affected directly by the monolayer. The authors asserted that the reduction in convective heat transfer could be explained, at least qualitatively, by a change in the hydrodynamic boundary condition at the surface due to the presence of a condensed monolayer. The evaporation resistance increased with increasing surface pressure of the monolayer.

The evaporation resistance-surface pressure curves for stearic acid exhibited bends which became sharp kinks near the freezing point of water. A possible explanation suggested by Navon and Fenn is the "squeezing out" of residual impurities in the monolayer.

EVAPORATION RATES OF FILM-COATED WATER DROPS

Snead and Zung[21] studied the evaporation rates of single liquid droplets coated with a film by means of a Millikan Oil Drop-type apparatus. The rates of evaporation were followed by observing changes in the voltage necessary to just suspend the droplet between the condenser plates as a function of time. Droplets of ethylene glycol coated with n-dibutylphthalate and droplets of water coated with n-decanol were the materials studied.

The radius of the coated water droplets was initially 1 to 5 microns (3.9×10^{-5} to 2.0×10^{-4} in.); the initial radius of the coated ethylene glycol droplets was somewhat larger. The relative humidity of the air was measured with a wet-dry bulb thermometer unit placed next to the Millikan apparatus.

The evaporation rates were given as the change of the square of the radius of the droplets, r^2, with time, t. This quantity was determined from the slope of the linear portion of a plot of r^2 against t. No values for the evaporation rates of pure water droplets were given because the droplets evaporated too rapidly in the apparatus for quantitative measurements to be made. The evaporation rate of the coated water droplets was as much as several hundred times slower than that of pure water droplets. In the case of coated ethylene glycol droplets the evaporation rate was lowered by only three to four times that of the pure droplets, indicating that the droplets were probably not completely coated.

The evidence obtained in this study on droplets coated with insoluble films indicates that the evaporation rates are always reduced by monolayers or films.

May[22] commented on the finding by Garrett[23] of a 9-fold reduction of the evaporation rate of water drops coated with a monolayer of cetyl alcohol and a 17-fold reduction with 1-docosanol, compared to the rate for a clean water drop. May studied the effects of films of $C_{18\text{-}22}H_{37\text{-}45}OC_2H_4OH$, OED, on the lifetime of water drops.

Water containing emulsified OED was sprayed onto fine spider threads wound across small rings. The rings were mounted in a device which allowed conditioned air to well upward over the droplets, simulating free-fall condition. Drops initially had diameters in the range of 10 to 150 microns (0.00039 to 0.0059 in.). The monolayer was rapidly completed and fully compressed.

Plots of drop diameter, d, against time, t, were excellent straight lines. During the life of a droplet, the straight line plot persisted until the last of the water was on the point of leaving the solid residue of monolayer substance. The direct proportionality between droplet lifetime and diameter was expressed in terms of minutes per micron of diameter lost, m. At 25°C and 50% relative humidity, m was 0.37 min/μm, representing an increase in lifetime of over 300-fold for a 30 micron (0.0012 in.) drop, for example. The value of m increased quite markedly with decreasing temperature so that, for example, at 2°C and 50% relative humidity, m was 30 min/μm. At higher temperatures, the product of m and saturation deficiency, $(1 - r)$, where r is relative humidity expressed as a decimal, was fairly constant.

May found that cetyl alcohol was much more effective in extending drop life than Garrett had suggested. Bent chain molecules, or those with large cross section, had no value in retarding evaporation. Studies were made only on drops which were supported by ultra-fine threads, 0.1 to 0.02 μm (4×10^{-6} to 8×10^{-7} in.).

Garrett[23] made a controlled laboratory investigation of the effects of chemical structure on the ability of surface films to retard water drop evaporation. Water drops were produced by the bursting of 2- to 3-mm (0.08- to 0.12-in.) diameter air bubbles at the surface of water contained in a glass-fritted bubbler. Pure monolayer-forming compounds spread at the air-water interface in the bubbler were effectively transferred onto the surfaces of the ejected jet drops. These jet drops, 100 to 500 μm (0.004 to 0.020 in.) in diameter, were collected on spider webs and/or Teflon™ fibers. Triply distilled water was used in all experiments except those involving fatty-acid films. The suspended drops were allowed to evaporate into open, quiescent laboratory air at 25 ± 0.2°C and 50 ± 2% or 30 ± 1% relative humidity. Fourteen different chemical systems were used to form monomolecular surface films.

Average evaporation rates of at least eight drops were determined from the slope of the plot of the square of the droplet radius against time. Expanded surface films produced little retardation of evaporation. The least permeable surface films, 1-docosanol and glycerol tristearate, produced a 17-fold decrease in evaporation rate over that of distilled water. Cetyl alcohol (1-hexadecanol) yielded a 9.1-fold reduction in evaporation rate at 25°C and 50% relative humidity. The presence of a small percentage of a branched-chain compound of the same carbon-chain length did not significantly reduce the effectiveness of an alkanol monolayer.

Extremely low evaporation rates for small drops result from measurement of the evaporation of the organic remnant of the protective film. This explanation was used to account for the reports of other workers of evaporation rate reductions by monolayers which were considerably greater than those determined by Garrett.

It may be concluded that there are sufficient nonlinear impurities in natural surface films of marine origin to reduce their ability to retard evaporation. Garrett's research established guidelines based upon chemical structure which could be used to judge the influence of natural surface films on water drop evaporation.

In replying to the comments of May, Garrett[24] repeated a previous conclusion of his "that some previously reported values of very large decreases in evaporation rate appear to be due to the inadvertent measurement of residual remnants of the organic film rather than the evaporation of water from a drop" and that he had "strongly suggested that the evaporation rate of the organic film-forming material was measured rather than that of coated water drops."

Garrett cited examples of experiments in which film-coated water drop diameters became constant and which, on inspection, were found to be empty shells of organic film-forming material. These observations were intended to provide a caveat against possible misinterpretation of evaporation data for film-coated drops. That is, one might be visually or instrumentally misled to the conclusion that a water drop is evaporating at an extremely slow rate, when in fact little or no water exists in the object under investigation.

Derjaguin et al.[25] studied the rate of evaporation of water drops after exposing them for a definite period of time to an atmosphere saturated with cetyl alcohol vapor. Prior to this work, interest in the retardation of evaporation of water due to the presence on the water surface of monolayers, primarily of cetyl alcohol, had increased in connection with the possibility of reducing evaporation losses from water reservoirs. Attempts had also been made to stabilize aqueous fogs by means of monolayers. No studies had been made of the possibility of applying insoluble films of low volatile substances to the surface of water by adsorption from the gas phase.

In this work, air saturated with cetyl alcohol was led into a thermostatically controlled cell in which a drop of radius 300 μm (0.0118 in.) was introduced on a glass filament of diameter 10 μm (0.00039 in.). The relative humidity was maintained at 100% so that the drop would not evaporate long enough for a cetyl alcohol adsorption layer to form. The flow was stopped, the drop was withdrawn from the cell, and its evaporation was observed under a microscope equipped with an ocular scale and an illuminator with a heat filter.

For the monolayer-coated drop, the evaporation was much slower than for the pure water drop, the lifetime of the coated drop increasing about 10-fold. The kinetics of the process also changed. So long as the monolayer was not saturated, its evaporation resistance was low compared with the resistance of air to diffusion; the presence of the unsaturated monolayer had no perceptible effect on the rate of evaporation. When the surface of the drop diminished to such an extent that the monolayer became saturated, evaporation became dependent mainly on the resistance of the monolayer.

The coefficient of condensation was found to equal 3.5×10^{-5}, which was in good agreement with available published data on the coefficient of condensation on a flat surface coated with a cetyl alcohol monolayer.

To obtain the adsorption isotherm, a study was made of the degree of saturation of the adsorption layer at various concentrations of cetyl alcohol vapor in the flow. The various concentrations were attained by diluting the air saturated with cetyl alcohol with a definite amount of filtered air free from alcohol vapor.

Derjaguin et al. concluded that on being adsorbed by the surface of a drop, cetyl alcohol vapor slows down its rate of evaporation considerably when the monolayer reaches saturation.

STABILIZATION OF WATER FOGS

Carlon and Shaffer[26] discussed the stabilization of water fogs against evaporation by the use of long-chain alcohol coatings. They reported on the retardation of evaporation of artificial water fogs by the application of cetyl alcohol (1-hexadecanol) to the droplets during generation.

Chamber and field tests were performed to obtain mass extinction coefficients for coated fogs produced by a commercial generator, the "FROSTOP."* The FROSTOP system consisted of a gasoline-driven air compressor and 6 to 10 generator units, each consisting of a water tub with a float-operated water inlet valve, a submersible propane burner, and ancillary hardware for connection of air, fuel, and makeup water lines to each generator unit. After the propane burner had been lighted, hot gases were carried by an air stream into the water where vigorous bubbling occurred, and the water was heated to an operating temperature of about 94°C. This temperature was high enough to melt cetyl alcohol bricks containing an emulsifer which were dropped into the generator tubs and quickly became uniformly mixed throughout the hot water bath.

The hot gases bubbling through the water carried heavy concentrations of water vapor and alcohol to the surfaces of the water baths. At the surface, cooling and condensation occurred and steam-like clouds of alcohol-coated water droplets became fully developed about 1 m (3.3 ft) above the tubs under typical atmospheric conditions.

The FROSTOP system was set up outdoors with one generator tub shrouded so that the water fog could be drawn directly into an aerosol optical test chamber for characterization. Measurements were made as the water fog dissipated over a 20- to 30-min period. Different cetyl alcohol concentrations determined fog droplet size and droplet extinction.

The two most significant observations were: (1) that while the alcohol coatings affect evaporation rate and droplet size, coated water fogs otherwise are similar spectrally to uncoated fogs; and (2) that droplet size is inversely proportional to the amount of coating material used, at least for the fogs produced by the FROSTOP systems. It was not possible to obtain definitive data on the evaporation rate of stabilized water fogs with relative humidity.

* The generator was marketed briefly by a subsidiary of the Boeing Aerospace Company under the trade name "FROSTOP."

It seems likely that evaporation-stabilized water fogs could be generated by systems smaller in scale than FROSTOP, perhaps using steam and alcohol mists or carrier streams to form droplets in predetermined size ranges.

The FROSTOP system was originally designed to generate evaporation-stabilized fogs to protect crops and orchards against freezing.

INHIBITION OF EVAPORATION FROM AGAR GEL

Roth and Loncin[27] developed a method for evaluating inhibition of evaporation, consisting of measuring the drying rate and the temperature of samples of agar in a stream of air. By treatment with an evaporation inhibitor, the decrease of drying rate and the increase of temperature allowed an accurate evaluation of inhibition.

It had been shown by Loncin[28] that molecular diffusion phenomena in a gel containing 2% agar and 98% water are practically the same as in pure water. It had also been shown that the rate of evaporation of water from a gel in a stream of air is exactly the same as the rate of evaporation of free water under the same conditions.

In the case of free water, the surface temperature is the wet-bulb temperature.

The experimental method consisted of dipping cylindrical gel samples of identical shape and size in aqueous suspensions or alcoholic solutions of surface active agents. By dipping a sample into a solution or suspension containing 0.1 to 1% of a suitable surfactant (e.g., hexadecanol or glycerol monostearate), evaporation is maximally reduced to 10% of the drying rate of an untreated sample.

The inhibition of evaporation was found to be extremely selective and limited mainly to some fatty alcohols and monoglycerides.

EFFECT OF MONOLAYERS ON RATE OF EVAPORATION OF WATER AND ON SOLUTION OF OXYGEN IN WATER

O'Brien et al.[29] studied the effect of monolayers of stearyl alcohol, stearic acid, oleic acid, dipalmitoyl phosphotidyl choline (saturated synthetic lecithin), and soybean lecithin on the rate of evaporation of water and on the solution of oxygen in water. The spreading solvent was hexane. Water evaporation and oxygen dissolution measurements were carried out in a Fabry-Perot interferometer.

In the water evaporation experiments, the interferometer was used as a thermometer to measure the temperature profile of the surface water layers. The magnitude of water evaporation retardation was inferred from an analysis of the interferograms. The percentage of retardation of evaporation by monolayers on water are for: (1) stearic alcohol, 59%; (2) stearic acid, 28%; (3) the lecithins, 20%; and (4) oleic acid, 5%.

In general, it appears that the area covered by a molecule in a compact monolayer is a good indication of the degree to which it retards evaporation of water. Retardation of oxygen penetration into water by the monolayers has the same general trend as evaporation retardation.

EFFECTS OF TRACES OF PERMEABLE SUBSTANCES

Langmuir and Schaefer[30] undertook experiments to test the hypothesis of the sensitivity of impermeable films to traces of permeable substances. The rate of weight increase of a drying agent, calcium chloride, supported on a wire gauze just above the surface of water in a trough was measured to determine the rate of evaporation of water.

The effect of compressed monolayers on the rate of evaporation was determined by making evaporation rate determinations first with a clean water surface and then with a surface covered by the monolayer.

A thermometer was placed in the water about 5 mm (0.2 in.) below the surface. Evaporation caused cooling of the water surface; during a 5-min period the temperature decreased by about 0.7°C. It was determined experimentally that the lowering of the water surface temperature was about 1.5 times that shown by the thermometer. The lowering of the water surface temperature decreased the saturation vapor pressure of the water and thus lowered the evaporation rate.

The absorption of the water by calcium chloride produced an increase in temperature. The resulting temperature gradient, due to the surface cooling and the absorption heating, had a marked effect in preventing convection currents and gave a linear distribution of partial pressure of water vapor in accord with the laws of diffusion. The resistance to evaporation, which may be considered to be proportional to the inverse of the rate of evaporation, was dealt with.

A series of measurements was made of the evaporation rate of water as a function of the distance from the water surface to the lower surface of the gauze supporting the calcium chloride. This distance was varied from 0.2 to 6.5 cm (0.08 to 2.6 in.). For values of distance, b, up to 2 cm (0.8 in.), a plot of evaporation resistance against b was a straight line. This indicated that the partial vapor pressure of water at the effective absorbing surface was constant and corresponded to a relative humidity of 21%. At b greater than 2, the evaporation resistance no longer increased linearly with b. This effect was attributed to convection due to minute temperature differences.

Langmuir and Schaefer concluded that: (1) the evaporation resistance is a property of condensed layers which are extremely sensitive to internal changes in the monolayers; (2) minute amounts of foreign substances may have great effects on the resistances; and (3) small amounts of a foreign substance can cause the collapse of a film that would otherwise withstand high pressure.

REDUCTION OF EVAPORATION FROM LAKES AND RESERVOIRS

Bean and Florey[31] studied the reduction of evaporation from Lake Hefner, Oklahoma by monomolecular films of mixtures of hexadecanol and octadecanol. A water sprinkler system introduced the film onto the lake. Measurements were made as the film spread under measuring instruments placed at 2 m (6.6 ft) above the lake surface.

The eddy-correlation technique was used to determine the rate of evaporation, E ($g/cm^2 \cdot sec$). E was calculated from

$$E = \overline{m'w'} \tag{2}$$

where m = the water vapor density (g/cm^3)
w = the vertical wind (cm/sec)

The fluctuations m′ and w′ are defined as $m' = (m - \overline{m})$ and $w' = (w - \overline{w})$. The overbar denotes a 10-min average of 1-sec observations. The 1-sec observations were the average of 30 observations during the 1-sec interval. The average of m′w′ was also determined for 10 min. The humidity was measured using a rapidly responding experimental barium fluoride element developed by Jones (see References 32 and 33) which was easily capable of resolving several cycles per second. The wind was measured with an anemometer-bivane.

The presence of the film affected both the amplitude and frequency of the humidity variations. The main effect was to suppress the magnitude of the short-term variations of humidity and thus to affect the magnitude of m′. The short-term fluctuations changed from film to no-film conditions about 20 to 30 min before the film either came under or left the water under the humidity sensor. This action reflected the amount of upwind water covered with the film.

The monomolecular film reduced evaporation. The average reduction in evaporation was 58%. This reduction during the 3-day test period was in excellent agreement with the values of 54 and 56% found in screening tests of the film-forming material. The screening tests were made in 4-ft-diameter evaporation pans under field conditions for an interval of 1 month.

Katsaros and Garrett[12] made the following observations concerning the practical effectiveness of monomolecular films in reducing evaporation from lakes and reservoirs.

> The practical use of these surfacant compounds to reduce evaporation from lakes and reservoirs was not always successful. The method of spreading the compound and the solvents used to disperse them modified the film coherence

while the surface waves and wind destroyed the films. The influences of circulation induced in the water by density differences and wind on evaporation retardation by monolayers are discussed in this paper. Limited success of programs to retard evaporation with organic films may be due in part to a misunderstanding of the forces involved, e.g., the author of ref. (2) (Jarvis[13]) says: 'A rather surprising result was noted for monolayers of cetyl alcohol and stearic acid when the air flow was about 6 L/min at 55% relative humidity (where there was considerable convection and the surface temperature was initially the same as the bulk water). The presence of a cetyl alcohol monolayer ... decreased the surface temperature 0.2 to 0.3°C, even though it reduced the rate of evaporation.'

Bouwer[34] made the following evaluation of the effectiveness of films:

Evaporation from lakes, reservoirs, or other water surfaces varies from about 2 m/yr for dry, hot climates to 1 m/yr or less for humid, cool climates. In the 1950's and 1960's, considerable research was done to reduce evaporation from open bodies of water by covering them with monomolecular layers of hexadecanol or octodecanol. While evaporation reductions of about 60 percent have been achieved under ideal conditions, actual reductions were much lower, and the use of monomolecular films to reduce evaporation from free water surfaces has found no practical application. Instead, more success has been obtained with floating objects like ping-pong balls, styrofoam blocks, or empty bottles. For tanks or other small reservoirs, floating sheets of foam rubber have been successfully used. Evaporation reduction of close to 100 percent have been obtained with such covers.

REFERENCES

1. Rideal, E. K. *J. Phys. Chem.* 29:1585 (1925).
2. La Mer, V. K., T. W. Healy, and L. A. G. Aylmore. "The Transport of Water Through Monolayers of Long-Chain n-Paraffinic Alcohols," *J. Colloid Sci.* 19:673—684 (1964).
3. Archer, R. J., and V. K. La Mer. "The Rate of Evaporation of Water Through Fatty Acid Monolayers," *J. Phys. Chem.* 59:200—208 (1955).
4. Shukla, R. N., S. B. Kulkarni, M. K. Gharpurey, and A. B. Biswas. "Retardation of Water Evaporation by the Monolayers of Octadecyl Alcohol and of Its Condensation Products with Ethylene Glycol," *J. Appl. Chem.* 14:236—239 (1964).
5. La Mer, V. K., and T. W. Healy. "Evaporation of Water: Its Retardation by Monolayers," *Science* 148:36—42 (1965).
6. Rosano, H. L., and V. K. La Mer. *J. Phys. Chem.* 60:348 (1956).

7. Chang, S. L., et al., *Retardation of Evaporation by Monolayers — Transport Processes* (New York, NY: Academic Press, 1962) p. 119.
8. Barnes, G. T., and V. K. La Mer. *Retardation of Evaporation by Monolayers — Transport Processes* (New York, NY: Academic Press, 1962) p. 9.
9. "Chemical Engineering Lab. Rept. No. SI-39," U.S. Bureau of Reclamation (1962).
10. MacRitchie, F. "Evaporation Retarded by Monolayers," *Science* 163:929—931 (1969).
11. MacRitchie, F. "Role of Monolayers in Retardation of Evaporation," *Nature* 218:669—670 (1968).
12. Katsaros, K. B., and W. D. Garrett. "Effects of Organic Surface Films on Evaporation and Thermal Structure of Water in Free and Forced Convection," *Intl. J. Heat Mass Transfer* 25:1661—1670 (1982).
13. Jarvis, N. L. "The Effect of Monomolecular Films on Surface Temperature and Convective Motion at the Water/Air Interface," *J. Colloid Sci.* 17:512—522 (1962).
14. Navon, U., and J. B. Fenn. "Interfacial Mass and Heat Transfer During Evaporation: I. An Experimental Technique and Some Results with a Clean Water Surface," *AIChE J.* 17:131—136 (1971).
15. Bartholic, J. F., J. R. Runkles, and E. B. Stenmark. "Effects of a Monolayer on Reservoir Temperature and Evaporation," *Water Resources Research* 3:173—179 (1967).
16. Sebba, F., and H. V. A. Briscoe. "The Evaporation of Water Through Unimolecular Films," *J. Chem. Soc. London* 106—114 (1940).
17. Cammenga, H. K., and D. Koenemann. "Continuous Recording of the Evaporation Rate of Water Through Monolayers," *Thin Solid Films* 159:53—61 (1988).
18. Navon, U., and J. B. Fenn. "Interfacial Mass and Heat Transfer During Evaporation: II. Effect of Monomolecular Films on Natural Convection in Water," *AIChE J.* 17:137—140 (1971).
19. Langmuir, I., and V. J. Schaefer. "Rates of Evaporation of Water Through Compressed Monolayers on Water," *J. Franklin Inst.* 235:119—162 (1943).
20. La Mer, V. K. *Retardation of Evaporation by Monolayers — Transport Processes* (New York, NY: Academic Press, 1962).
21. Snead, C. C., and J. T. Zung. "The Effects of Insoluble Films Upon the Evaporation Kinetics of Liquid Droplets," *J. Colloid Interface Sci.* 27:25—31 (1968).
22. May, K. R. "Comments on 'Retardation of Water Drop Evaporation with Monomolecular Surface Films'," *J. Atmos. Sci.* 29:784—785 (1972).
23. Garrett, W. D. "Retardation of Water Drop Evaporation with Monomolecular Surface Films," *J. Atmos. Sci.* 28:816—819 (1971).
24. Garrett, W. D. *J. Atmos. Sci.* 29:786—787 (1972).
25. Derjaguin, B. V., V. A. Fedoseyev, and L. A. Rosenzweig. "Investigation of the Adsorption of Cetyl Alcohol Vapor and the Effect of This Phenomenon on the Evaporation of Water Drops," *J. Colloid Interface Sci.* 22:45—50 (1960).

26. Carlon, H. R., and R. E. Shaffer. "Optical Properties (0.63—13 μm) of Water Fogs Stabilized Against Evaporation by Long-Chain Alcohol Coatings," *J. Colloid Interface Sci.* 82:203—207 (1981).
27. Roth, T., and M. Loncin. "Evaluation of Water Evaporation Inhibitors," *J. Colloid Interface Sci.* 100:216—219 (1984).
28. Loncin, M. *Die Grundlagen der Verfahrenstechnik in der Lebensmittelindustrie* (Aarau: Verlag Sauerländer, 1969).
29. O'Brien, R. N., A. I. Feher, K. L. Li, and W. C. Tan. "The Effect of Monolayers on the Rate of Evaporation of H_2O and the Solution of O_2 in H_2O," *Can. J. Chem.* 54:2739—2744 (1976).
30. Langmuir, I., and V. J. Schaefer. "Rates of Evaporation of Water Through Compressed Monolayers on Water," *J. Franklin Inst.* 235:119—162 (1943).
31. Bean, B. R., and Q. L. Florey. "A Field Study of the Effectiveness of Fatty Alcohol Mixtures as Evaporation Reducing Monomolecular Films," *Water Resour. Res.* 4:206—208 (1968).
32. Jones, F. E., and A. Wexler. "A Barium Fluoride Film Hygrometer Element," *J. Geophys. Res.* 65:2087—2095 (1960).
33. Bean, B. R., and R. O. Gilmer. "Comparison of Barium Fluoride Humidity Element with the Microwave Refractometer for Studies of Rapid Fluctuations of Atmospheric Humidity," *Radio Science* 4:1155—1157 (1969).
34. Bouwer, H. "Water Conservation," *Agri. Water Manage.* 14:233—241 (1988).

CHAPTER 9

Equations Used to Calculate Evaporation Rate and Evapotranspiration

INTRODUCTION

A variety of equations are used to calculate evaporation rate and evapotranspiration. In this chapter, applications of these in experimental situations are made.

BULK AERODYNAMIC EQUATION

Conaway and van Bavel[1] made a direct evaluation of the theoretical mass transfer equation (also referred to as a bulk aerodynamic equation or a Dalton-type expression) used to calculate evaporation from open water. The equation can be written as

$$E = (\rho r/p)(e_o - e_z)f(u_z) \quad (1)$$

where E = evaporation rate (in g/cm²·sec)
ρ = density of air
r = ratio of the molecular weight of water to the molecular weight of air
p = ambient pressure
e_0 = vapor pressure of water at the water surface
e_z = vapor pressure of water at elevation z
f = wind function
u_z = wind speed at elevation z

A simple form of the wind function[2] is

$$f(u_z) = k^2u_z/[\ln(z/z_o)]^2 = C_zu_z \quad (2)$$

where k = Von Kármán constant
z_0 = surface roughness parameter
C_z = drag coefficient

This equation was referred to as the log-law wind function.

The surface temperature was measured using a narrow bandpass infrared radiation thermometer, with an accuracy of $\pm 0.2°C$. Water vapor pressure in air and wind speed over a wet bare surface were also measured. The calculated values of evaporation rate were compared with evaporation rate measured using weighable lysimeters.

Field measurements were carried out on a 70 × 90 m (230 × 295 ft) experimental area. The field was level and free of plants. It was irrigated with 12 cm (4.7 in.) of water on an evening and was visibly wet throughout the following two days. The weather was warm and fair, with occasional scattered clouds. Evaporation was determined from 15-min triplicate weighings of three 1-m^2 (11-ft^2) lysimeters located centrally in the field. The accuracy of the lysimeters was ±0.02 mm (0.0008 in.) of water.

The data showed that under the conditions of the experiment, the log law used in a Dalton-type expression gave accurate values of the evaporation from a wet bare surface. The findings also suggested that remote sensing of surface radiation from a flying observatory could be applied to the calculation of evaporation, when combined with surface information on air humidity, wind speed, surface emitance, and estimate of sky radiance.

THE PENMAN EQUATION

The Penman[3] equation for estimating potential evapotranspiration, E_0, can be written as:

$$E_0 = \frac{(\Delta/\gamma)\, H_e + E_a}{(\Delta/\gamma) + 1} \tag{3}$$

where
- Δ = slope of the saturation absolute humidity curve at the air temperature
- γ = the psychrometric constant
- H_e = evaporation equivalent of the net radiation
- E_a = aerodynamic expression for evaporation

Alternatively in Brutsaert's[4] derivation, Q_{ne} is substituted for H_e and E_A is substituted for E_a. Q_{ne} is the ratio of the available energy flux density to the latent heat of vaporization, and E_A is a drying power of the air. In the Brutsaert derivation, Δ is the slope of the saturation water vapor pressure curve.

To take into account the difference between air temperature and water temperature in computing radiation emitted from a water surface, Kohler and Parmele[5] modified the Penman[3] equation

$$E = (Q_n\Delta + E_a\gamma)/(\Delta + \gamma) \tag{4}$$

where
- E = evaporation rate (in./day)
- Q_n = net radiation (erg/cm^2-sec)
- Δ = slope of the saturation water vapor pressure versus temperature curve at the air temperature (T_a, in °K)

E_a = evaporation rate defined by $E_a \equiv f(u)(e_s - e_a)$ assuming that the surface temperature (T_0, in °K) is equal to the air temperature
$f(u)$ = wind function (u is wind velocity, cm/sec) appearing in a form of Dalton's equation
e_s = saturation water vapor pressure at the air temperature
e_a = water vapor pressure of air at height a above the surface
γ = psychrometric constant

The approach taken was to express Q_n as

$$Q_n = Q_{ir} - Q_b = Q_{ir} - \epsilon\sigma T_0^4 \tag{5}$$

where Q_{ir} = difference between incident and reflected radiation (all-wave)
Q_b = emitted radiation
ϵ = emissivity of the water surface
σ = Stefan-Boltzmann constant

Letting $(T_0 - T_a) = (e_0 - e_s)/\Delta$ and substituting for Q_n using the first two terms of a binomial expansion of $[T_a + (T_0 - T_4)]$,[4]

$$E = \frac{(Q_{ir} - \epsilon\sigma T_a^4)\,\Delta + E_a\,[\gamma + 4\epsilon\sigma\, T_a^3/f(u)]}{\Delta + [\gamma + 4\epsilon\sigma\, T_a^3/f(u)]} \tag{6}$$

E_a was computed for generalized lake evaporation estimates using the equation

$$E_b = (0.181 + 0.00236\, u_4)(e_s - e_2) \tag{7}$$

where u_4 = wind velocity at an elevation of 4 m, observations taken over land
e_2 = water vapor pressure at an elevation of 2 m, observations taken over land

Account is taken of advected energy and changes in energy storage by the term $\phi[Q_i - Q_0 - (S_2 - S_1)]$

where ϕ = $\Delta/[\Delta + \gamma + 4\epsilon\sigma T_0^3/f(u)]$
Q_i = energy content of the inflow of water
Q_0 = energy content of the outflow
$(S_2 - S_1)$ = increase in energy storage during the period of interest

Combining Equation 6 with $\phi[Q_i - Q_0 - (S_2 - S_1)]$,

$$E_L = \phi[Q_i - Q_0 - (S_2 - S_1)] \tag{8}$$

where E = computed free-water evaporation assuming "net advection" $[Q_i - Q_0 - (S_2 - S_1)]$ to be equal to zero
E_L = estimated evaporation from specific lakes

Equations 6, 7, and 8 were tested using data from Lake Hefner, OK and Felt Lake, CA (water-budget measurements); Lake Meade, AZ-NV (energy-budget and mass transfer estimates); Silver Hill, MD (15-ft-diam sunken evaporation pan); and Sterling, VA (16-ft-diam evaporation pan). The results of the comparison for Lake Meade showed the largest bias, 18%, significantly larger than the bias in each of the other cases.

The authors concluded that, based on the limited verification, when appropriate data are available: Equation 6 provides a suitable basis for generalized estimates of mean annual free-water evaporation; Equation 6 in conjunction with Equation 8 should provide reliable estimates of monthly evaporation from existing reservoirs; and Equation 6 might be used to estimate daily potential evapotranspiration for application to hydrologic models.

EVAPOTRANSPIRATION DETERMINATION BY THE BOWEN RATIO METHOD

Tanner,[6] in a humid region, and Fritschen,[7] in a warm arid region, had established the accuracy of the Bowen ratio method in determining evapotranspiration rates. The absolute accuracy was about 95% when compared with lysimeter measurements. Fritschen[8] then undertook to illustrate the usefulness of the method as a survey technique or for continuous evaporation determinations.

By the Bowen ratio method the evaporative flux is given by

$$LE = \frac{-(R_n + S)}{1 + (C_pP/\epsilon L)(K_h/K_w)(\overline{\Delta T}/\overline{\Delta e})} \tag{9}$$

where LE = evaporative flux in langleys (cal/cm^2)
R_n = net radiation in langleys
S = soil heat flux in langleys
C_P = specific heat of dry air at constant pressure
P = atmospheric pressure
L = the latent heat or vaporization of water
ϵ = ratio of the molecular weight of water vapor to that of dry air
K_h and K_w = transfer coefficients of sensible heat and water vapor, respectively

$\overline{\Delta T}$ and $\overline{\Delta e}$ = average air temperature and water vapor pressure differences, respectively, between two heights above the crops of 5 and 40 cm (2.0 and 16 in.).

LE, R_n, and S are more appropriately expressed as langleys/min. The Bowen ratio, a ratio of the similarity equations for sensible and latent heat, is the term beyond the 1 in the denominator of the equation.

Evapotranspiration determinations using the Bowen ratio were made simultaneously over the following crop combinations: alfalfa and barley, alfalfa and cotton, alfalfa and sorghum, wheat and oats, and cotton. Adequate soil moisture was maintained in the experimental fields by frequent irrigation.

Data were collected for two or three 24-hr periods (midnight to midnight), the meteorological sensors and a mobile laboratory were moved to another location with different crops. The various crop combinations were sampled twice monthly, a total of 37 runs were made from February to September at Mesa and Phoenix, AZ. The Bowen ratio equation was solved every 15 min, using sampled values of net radiation and soil heat flux and 15-min moving averages of air-temperature and vapor-pressure differences. The moving averages were used, in conjunction with the reversing of sensors, to cancel out any instrumental bias prior to computing the Bowen ratio.

The experimental results showed that evapotranspiration generally exceeded net radiation, indicating that energy was derived from the air mass. Ratios of evapotranspiration to net radiation ranged from 1.0 to 1.8. Fritschen concluded that "the Bowen ratio method, as a survey technique or continuous sampling method, can be used to obtain short-period evapotranspiration rates under field conditions. It appears to be one of the few methods which will yield valid results when crops are subjected to frequent and heavy irrigation or where water tables are present."

COMPARISON OF EQUATIONS

Warnaka and Pochop[9] stated that a relatively small amount of information existed on the ability of various equations to define the variation of evaporative losses. An immediate need for their analysis was to select an evaporation equation which would permit calculation of long-term records of monthly and annual evaporation. It was critical that estimated evaporation have variability characteristics similar to those of actual evaporation, for the calculation of probabilities from the results of a routing procedure to analyze the water balance of evaporation ponds and thus the probabilities of success or failure of specific pond designs. A major consideration was how well any equation describes the variation within evaporation rates.

Estimates from six climatological methods for predicting free-water evaporation were compared. The equations included: (1) the Kohler-Nordenson-Fox; (2) the Kohler-Parmele; (3) the Linacre; (4) the Priestley-Taylor; (5) the

Stewart-Rouse; and (6) the deBruin equations. All of these equations have a theoretical formulation based on Penman's derivations. Monthly and total estimates of free-water evaporation were made for each of the six equations and five months at Pathfinder Dam and Whalen Dam in Wyoming. These two locations were chosen because of the availability of Class A evaporation pan records of considerable length.

To predict free-water evaporation from ponds, the pan readings were multiplied by a pan coefficient of 0.7. The only climatological data at the pan stations were air temperatures. Means and standard deviations of the monthly values were calculated and t-tests and regressions comparing the data from the various equations and against pan data were performed. The comparison of the six equations showed that the equations varied greatly in their ability to define the magnitude and variability of evaporation.

EDDY CORRELATION

A quotation from the World Meteorological Organization Technical Note No. 83 (Measurement and Estimation of Evaporation and Evapotranspiration)[10] will serve to introduce this section: "Currently the only truly direct method for obtaining rate of evaporation is the eddy-correlation method."

The eddy-correlation method as applied to water-vapor transport is a direct method for determining the vertical transfer of water vapor by eddies in the lower atmosphere.[11] The time average of the water-vapor flux at a fixed point is given by

$$\overline{\rho w\, q} = \overline{\rho w}\; \overline{q} + \overline{(\rho w)'q'} \tag{10}$$

where water vapor flux can be expressed in $g/cm^2 \cdot sec$, and

where ρ = air density (g/cm^3)
w = vertical velocity (cm/sec)
q = specific humidity defined by $q \equiv 0.622\ e/p$, where 0.622 is the ratio of the molecular weight of water to the mean molecular weight of dry air
e = water vapor pressure
p = total or atmospheric pressure

The overbar indicates time average and the prime indicates instantaneous departure from the mean.

According to Priestly,[12] "At a low height and over a period long enough to embrace a large number of eddy passages, w will become small unless there is a very large horizontal convergence or divergence below the point considered, and this can generally be excluded where the terrain is uniform."

It will be helpful to define "eddy" through a quote from Eady:[13] "we may think of air motion as composed of superimposed flow patterns of different 'scales'. Since the motion often appears to be rather like the circular motion in an eddy, we may rather loosely refer to the patterns, or parts of patterns, as 'eddies'. The scale can then be defined as the average horizontal size of the eddies." Eddies occur in all sizes; a continuous hierarchy exists from the largest (limited by the dimensions of the fluid system) to the smallest eddies.[14]

The last term in Equation 10 is the turbulent or eddy flux of water vapor ($g/cm^2 \cdot s$) and is equal to the rate of evaporation of water. The eddy flux of latent heat is the product of the latent heat of vaporization of water and the last term in Equation 10. The direct measurement of the eddy flux of water vapor provides a direct measurement of the rate of evaporation of water.

Measurements of evaporation rate by the eddy-correlation method[15,16] have been limited by the relatively slow response of the humidity sensors. However, the barium fluoride film humidity sensor developed by Jones[17] has been used in conjunction with a three-dimensional anemometer for evaporation measurements over a snap bean crop.[18]

Goltz et al.[18] made evaporation measurements by an eddy correlation method. They considered, among the micrometeorological methods available at that time for determining evaporation, the eddy correlation approach to be most satisfying since it required the least number of basic assumptions.

The equation describing the evaporation as latent flux density was written as

$$F_v = \lambda[\langle q \rangle \langle w \rangle + \langle q'w' \rangle] \tag{11}$$

where λ = latent heat of vaporization of water
q = vapor concentration (absolute humidity)
w = vertical wind velocity

The angle brackets indicate time averages, and the primes indicate fluctuations about the mean. The surface evaporation is expressed as

$$E = \lambda \langle q'w' \rangle \tag{12}$$

when the time average of the vertical wind velocity is zero.

In order to work closer to the ground for which rapidly responding sensors are needed, Goltz et al.[18] investigated the possible use of the barium fluoride-film humidity sensor.[17] Bean and Florey[19] had reported on the use of this sensor for measuring evaporation at Lake Hefner. Since the system at Lake Hefner was limited by the relatively slow response of an anemometer bivane and not by the humidity element's response, Goltz et al. believed that the barium fluoride-film humidity sensor, in association with a fast-response wind-vector sensor, could allow measurement of evaporation considerably closer to the surface.

EVAPORATION FROM A ROUGH SURFACE

Brutsaert[20] analyzed evaporation from a rough surface as a phenomenon of molecular diffusion into random-lived internal scale turbulent eddies, using a model based on the principles of similarity and stochastic surface renewal. He made the following general statement concerning natural surfaces:

> Conditions in nature are usually such that the air flow just above it is turbulent. Turbulent flow may conveniently be considered as consisting of a mean motion on which a secondary fluctuating motion is superimposed. This secondary motion consists of random oscillations and rotations of particles, or eddies; their size, usually referred to as scale, covers a very wide range. The two extreme sizes of eddies are referred to as the external and internal scale of turbulence. Throughout most of this range the smaller eddies obtain energy by inertial transfer from the larger ones.

The length scale of the smallest eddies (internal length scale) in the immediate vicinity of the surface is given by

$$\ell = (\nu^3/\epsilon)^{1/4} \qquad (13)$$

where ν = kinematic viscosity of the fluid
ϵ = rate of dissipation of turbulent energy in the flow

ϵ becomes

$$\epsilon = u_*^3/k_v z_0 \qquad (14)$$

where u_* = $(\tau/\rho)^{1/2}$ friction velocity
τ = sheer stress
ρ = density of the air
k_v = von Karman's constant
z_o = roughness length

The friction velocity takes into account the stability of the atmosphere and under neutral conditions $u_* = u$, the average wind speed at elevation z. ℓ then becomes

$$\ell = (\nu^3 k_v z_0/u_*^3)^{1/4} \qquad (15)$$

and the internal time scale

$$\Theta = (\nu/\epsilon)^{1/2} = (\nu k_v z_0/u_*^3)^{1/2} \qquad (16)$$

ℓ is of the order of mm to cm[12] in the atmosphere. The total rate of evaporation from the entire surface is

$$E = -\rho D s_e \underline{L}\{(\partial q/\partial z)_{z=0}\} \tag{17}$$

where E = rate of evaporation (g/cm²·sec)
D = molecular diffusivity
s_e = average renewal rate of the eddies in contact with surface
$\underline{L}$ = Laplace transform
$(\partial q/\partial z)_{z=0}$ = specific humidity gradient at the surface

Under Dankwerts'[21] assumption that molecular diffusion from the surface into the eddies may be considered as diffusion into a stagnant fluid of infinite depth, the appropriate boundary conditions are: $q = q_0$, $0 < z < \infty$, $t = 0$; $q = q_0$, $z = 0$, $t \geq 0$; $q = q_0$, $z = \infty$, $t \geq 0$. Under Dobbins'[22] assumption that diffusion into the eddy is similar to diffusion into a stagnant film of effective thickness, $A\ell$, the appropriate boundary conditions are: $q = q_a$, $0 < z < A\ell$, $t = 0$; $q = q_0$, $z = 0$, $t \geq 0$; $q = q_a$, $z = A\ell$, $t \geq 0$.

The rate of evaporation from a large wet surface into turbulent air becomes

$$E = C\rho D^{1/2}(q_0 - q_a)u_*^{3/4}/(\nu k_v z_0)^{1/4} \tag{18}$$

where C is to be determined by experiment
q_0 = specific humidity of saturated air at the temperature of the surface
q_a = specific humidity in the main body of a fully turbulent, perfectly mixed air stream

It should be stressed here that it is assumed that the air stream is fully turbulent and perfectly mixed so that the water vapor concentration in the turbulent air is constant and uniform with elevation. Indicating the importance of the appearance of $u_*^{3/4}$ and $D^{1/2}$ in Equation 18, Brutsaert[20] compared the values of the exponents to values found experimentally in wind tunnels and quoted the following values for the exponent of velocity: Shepherd et al.,[23] 0.75; Millar,[24] 0.725 and 0.77; Himus,[25] approximately 0.77; Powell and Griffiths,[26] 0.85; and Gililand and Sherwood,[27] for vaporization into turbulent air inside wetted wall pipes, 0.83. Brutsaert attributed the experimental values of the exponent of velocity to measurements in a transitional flow regime between fully rough and smooth. Experimental values for the exponent of D cited included: Gililand and Sherwood,[27] 0.56; Deissler,[28] between 0.5 for low Schmidt (or Prandtl) numbers and 0.75 for very high Schmidt (or Prandtl) numbers.

Brutsaert concluded that the theory developed in his paper could only be applied to evaporating surfaces that are fairly regular. Also that since the model is one-dimensional, it applies strictly to surfaces of infinite extent; however, by means of appropriate coefficients (as with evaporation from small pans) this difficulty might be overcome.

Calculated Evapotranspiration from Remotely Sensed Reflected Solar Radiation and Surface Temperatures with Ground-Station Data

Reginato et al.[29] reported work intended to demonstrate that remotely sensed multispectral data (visible, near infrared, and thermal infrared) could be combined with ground-based meteorological data to evaluate evapotranspiration and that the "remote" values compare favorably with data obtained from weighing lysimeters and from soil-content changes.

Determination of evapotranspiration from remotely sensed observations requires evaluation of net radiation, soil heat flux, and sensible heat flux, as given by the energy balance for the surface:

$$LE = R_n - G - H \qquad (19)$$

where in units of w/m^2,

LE = latent heat flux
R_n = net radiation
G = soil heat flux
H = sensible heat flux

Net radiation can be written as the sum of incoming and outgoing radiation. Jackson et al.[30] argued that the incoming terms could be measured with traditional ground-based instruments and the data extrapolated radially for some distance from the point of measurement. They also showed that the outgoing terms could be evaluated from multispectral data, and suggested that evaluating the terms in this manner would allow the construction of detailed net radiation maps over relatively large areas, up to tens of square kilometers (multiples of four square miles). The authors assumed that the soil heat flux could be adequately estimated by a linear function of crop height. The empirical relation was

$$G = (0.1 - 0.042h)R_n \qquad (20)$$

where h = canopy height in m, assumed to be 1.2 m (3.9 ft) at maturity

The sensible heat flux, H, can be expressed as

$$H = \rho c_p(T_s - T_a)/r_{ac} \tag{21}$$

where ρc_p = volumetric heat capacity (although it is slightly temperature-dependent, they used a constant value of 1200 $J/m^3 \cdot °C$)
T_s = surface temperature
T_a = air temperature at height z (in meters) above the surface
r_{ac} = stability-corrected aerodynamic resistance, which depends on a number of parameters including the roughness length, z_0, and h

They considered z_0 to be closer to 0.05 h than the conventional 0.13 h for a developed wheat canopy.

Using the relationships above, the evapotranspiration, ET, actually the latent heat flux, LE, was calculated as

$$LE = (0.9 + 0.042h)R_n - 1200(T_s - T_a)/r_{ac} \tag{22}$$

Calculation of evapotranspiration by combining remote and ground-based measurements is based on the fact that the latter can be extrapolated to areas larger than the immediate surroundings. Calculations were made of the error in ET that might result from errors associated with extrapolation and with measurement. These calculations indicated that errors in measurement or extrapolation of air temperatures and wind speed under clear-sky conditions have significant effect on the calculation of evapotranspiration. This raised a question regarding the density of the network of measurements that is required to obtain adequate estimates of ET. The calculations also indicated the need for accurate values of the roughness length, z_0.

For 18 wheat plots for 44 cloudless days over a growing season, instantaneous values of ET were calculated. Lysimeters in 3 of the 18 plots provided data to compare against the instantaneous values. For the remaining plots, daily ET was estimated from the instantaneous data and compared with values calculated from soil moisture contents measured using a neutron moisture meter. The comparisons indicated that, for generally clear-sky conditions, ET could be adequately evaluated using a combination of remotely sensed and ground-based meteorological data. The results suggested that ET maps of relatively large areas could be made using this method with data from airborne sensors. The extent of the area covered appeared to be limited by the distance that air temperature and windspeed data could be extrapolated.

Evaluation of Evaporation Using Airborne Radiometry and Ground-Based Meteorological Data

Jackson et al.[31] combined airborne measurements of reflected solar and emitted thermal radiation with ground-based measurements of incoming solar radiation, air temperature, windspeed, and vapor pressure to calculate instantaneous evaporation (LE) rates using a form of the Penman equation. Estimates of evaporation on five days during a one-year period were made for cotton, wheat, and alfalfa fields. Simultaneous use of a Bowen ratio apparatus provided ground-based measurements of evaporation.

There was good agreement between the airborne and ground-based techniques, with a greatest difference of about 12% for the instantaneous values. Estimates of daily evaporation were made from the instantaneous data. On three of the five days, the difference between the two techniques was less than 8%, with the greatest difference being 25%.

The results demonstrated that airborne remote sensing techniques can be used to obtain spatially distributed values of evaporation over agricultural fields.

Estimating Evaporation Using Thermal Infrared Satellite Imagery

Miller and Millis[32] studied the use of satellite thermal infrared data to aid in the estimation of evaporation from lakes and reservoirs and evaluated the effects of spatially varied salinity over the Great Salt Lake. The study was conducted on Utah's Great Salt Lake because of the availability of both satellite and meteorological data.

Pan-evaporation data from five local weather stations were used. The pan-evaporation values were multiplied by a pan coefficient of 0.7 to obtain equivalent freshwater-lake evaporation. Pan-saltwater evaporation values were obtained by multiplying pan data by salt coefficients, these values were then multiplied by the pan coefficient to obtain saltwater-lake evaporation values. An average satellite thermal intensity value was converted to an average water surface temperature.

A surface temperature-evaporation model that considered the large areal extent of the lake, 1,500,000 acres (610,000 ha), and its varying salt concentrations were used. The surface temperature-evaporation relationships and the correlation coefficients of the data were determined by linear regression.

The authors considered the correlation results to be generally very good and that their results indicated that models can be developed to estimate evaporation from the entire lake and from smaller sections of the lake while taking into account the salinity of the lake.

Evaporation from Heated Water Bodies

Chattree and Sengupta[33] undertook a study to develop methods for predicting net heat flux and evaporation across the air-water interface, and made experimental measurements at a cooling pond at East Mesa, CA. The difference, ΔT, between bulk water temperature and water surface temperature had a broad range of values from 0.1 to 2°C. The two temperatures were measured by the same instrument. The water temperature varied from 30 to

46°C, the air temperature varied from 22 to 44°C. The relative humidity varied from 28 to 68%. The wind speed at a height of 4.5 m (15 ft) above the water surface was in the range 0.7 to 5.8 m/sec (2.3 to 19 ft/sec). The evaporative heat flux, a major component of the net heat flux, could be accurately known since the pond bottom had a plastic lining and there was no rain during the experiments.

The elevation of the pond surface was determined with respect to an arbitrary reference level by reading water levels in stilling wells using hook gauges. Total downward solar radiation was measured with an Eppley precision pyranometer located within 10 m (33 ft) of the edges of the pond, net radiation for all wavelengths was measured over the pond surface with a dual-hemisphere Fritschen-type radiometer. A radiometer was used to measure the surface temperature of the water, no thermal stratification was observed in the horizontal direction. Surface temperature readings were averaged for 2-min intervals. The bulk liquid temperature was measured by the radiometer, and averaged for 10-sec intervals after the surface was broken by a jet from a pump submerged about 20 cm (7.9 in.) below the water surface. Air temperatures and humidities were measured using aspirated bead thermistors, dry and wet wicks. Wind velocity and wind direction were measured at a reference tower using three orthogonal propellers at each height.

The evaporative heat flux was calculated by finding the difference between the averaged stilling well readings at time intervals of 1 hr. An analysis was made of 70 data points covering both night and day conditions. The net heat flux at the interface was out of the surface and was negative at all times. The surface temperature was lower than the bulk temperature at all times. The response of the surface temperature to a change in meteorological conditions was of the order of minutes or perhaps seconds, while the major component of the net heat flux was the evaporative heat flux which was averaged over an hour.

An attempt was made to predict the net heat flux and the evaporative heat flux crossing the surface of a heated water body. Existing formulas which used the surface temperature difference and wind speeds to predict net heat flux were examined. A dimensional correlation which includes the effect of surface temperature variations was found to give the most accurate prediction of the net heat flux.

Evaporation from heated water bodies was studied by Adams et al.,[34] using data collected at the National Geothermal Test Facility at East Mesa, CA and at the Savannah River Site near Aiken, SC.

The first study was conducted on an exposed, heated, 0.3-ha (0.7-acre) lined cooling pond at East Mesa. The major objective of this study was to acquire high-quality data using conventional water and energy budget approaches. The second study was made on a series of ponds and interconnecting channels at the Savannah River Site, at which data under extreme water temperatures of about 70°C were obtained. At these temperatures, free convection dominated surface heat exchange; evaporation was determined primarily using the energy budget approach. At both sites, alternate measurement techniques were studied which might be used to measure evaporation, including vapor budget, immersed pans, and correlation with surface (skin) temperature difference.

Evaporative heat loss from a body of water is attributed to: (1) forced convection due to the horizontal movement of air across the water surface; and (2) free convection due to the density difference between air at the water surface and in the ambient surroundings. Both components contribute significantly to evaporation from a cooling pond. In attempting to predict evaporation from heated water bodies, an understanding of how the free and forced components of evaporative heat loss add is necessary.

A total of six "experiments" was performed at the East Mesa site; consisting of filling the pond (depth 1.5 m, 4.9 ft) with geothermal fluid and monitoring for about 48 hr the water surface elevation, water temperature, wet- and dry-bulb air temperature, net radiation, incident solar radiation, wind speed and direction, ground temperatures, and barometric pressure. The time-varying water level and temperature (to account for thermal expansion or contraction of the water) were used to calculate water budget evaporation. At the Savannah River Site (to observe the effects of the warmest possible temperatures), experiments were concentrated on the channel leading directly from a nuclear reactor and on the first precooler pond. The channel dimensions were approximately 2500 m (8200 ft) × 13 m (43 ft) × 0.5 m (1.6 ft) deep. The pond was approximately square with an area of about 4 ha (10 acres) and with depth ranging from 1 to 2.5 m (3.3 to 8.2 ft). Additional measurements were made on a downstream precooler pond with surface area of about 50 ha (124 acres). Using the observed differences in inflow and outflow temperature, evaporative heat loss was calculated by applying dynamic energy budgets to the water bodies.

Evaporative heat loss due to free convection was treated as analogous to heat loss from a horizontal heated plate. The overall mass and heat transfer was affected by a thin laminar sublayer above the air-water interface, of the order of 1 mm (0.04 in) in thickness. The temperature drop between the bulk of the water and the surface ranged from 0.1 to 2.3°C, with an average of about 0.8°C, at the East Mesa site; at the Savannah River Site, it ranged between 0.2 and 3.7°C, with an average of about 1.8°C. Prior to these investigations, the water temperature of most lakes and cooling ponds studied ranged between 20 and 30°C. At East Mesa, the water surface temperature varied between 35 and 40°C; at the Savannah River Site, it varied between 58 and 73°C. Under Savannah River Site conditions, the free evaporation coefficient was reduced from the corresponding value at a water surface temperature of about 25°C by 18%. The areal average evaporation from a pond is reduced as the size (fetch) of the pond increases, since the humidity increases with downwind distance over the pond.

Evaporation at the East Mesa site was more efficient than that at the Savannah River Site due to its exposure in a desert setting. The effective rate of evaporation at the Savannah River Site was decreased since the moisture was easily trapped by the trees in the vicinity of the channel and ponds. A new equation was developed that combines free and forced convection in a different manner from previously developed equations, evaporation was computed as the square root of the sum of the squares of the respective components. Pond fetch was shown to be a significant factor which increases the forced convection component of evaporation in small exposed ponds, such as at East Mesa.

REFERENCES

1. Conaway, J., and C. H. M. van Bavel. "Evaporation from a Wet Soil Surface Calculated from Radiometrically Determined Surface Temperatures," *J. Appl. Meteorol.* 6:650—655 (1967).
2. Sverdup, H. V. "The Humidity Gradient Over the Sea Surface," *J. Meteorol.* 3:1—8 (1946).
3. Penman, H. L. "Natural Evaporation from Open Water, Bare Soil, and Grass," *Proc. Roy. Soc. London* A193:120—146 (1948).
4. Brutsaert, W. "A Model for Evaporation as a Molecular Diffusion Process into a Turbulent Atmosphere," *J. Geophys. Res.* 70:5017 (1965).
5. Kohler, M. A., and L. H. Parmele. "Generalized Estimates of Free-Water Evaporation," *Water Resour. Res.* 3:997 (1967).
6. Tanner, C. B. "Energy Balance Approach to Evapotranspiration from Crops," *Soil Sci. Soc. Am. Proc.* 24:1—9 (1960).
7. Fritschen, L. J. "Accuracy of Evapotranspiration Determination by the Bowen Ratio Method," *Bull. Intern. Assoc. Sci. Hydrol.* 10:38—48 (1965).
8. Fritschen, L. J. "Evapotranspiration Rates of Field Crops Determined by the Bowen Ratio Method, *Agron. J.* 58:339—342 (1966).
9. Warnaka, K., and L. Pochop. "Analyses of Equations for Free Water Evaporation Estimates," *J. Geophys. Res.* 93:979—984 (1988).
10. *Measurement and Estimation of Evaporation and Evapotranspiration* (Geneva, Switzerland: Secretariat of the World Meteorological Organization, 1966) World Meterological Organization Tech. Note No. 83.
11. Swinbank, W. C. "The Measurement of Vertical Transfer of Heat and Water Vapor by Eddies in the Lower Atmosphere," *J. Meteorol.* 8:135 (1951).
12. Priestley, C. H. B. *Turbulent Transfer in the Lower Atmosphere* (Chicago, IL: University of Chicago Press, 1959), p. 4.

13. Eady, E. T. "The General Circulation of the Atmosphere and Oceans," in *The Earth and Its Atmosphere*, D. R. Bates, Ed. (New York: Basic Books, Inc., 1957), p. 131.
14. Hess, S. L. *Introduction to Theoretical Meteorology* (New York: Holt, Rinehart, and Winston, 1959), p. 267.
15. Dyer, A. J., and F. K. Maher. "Automatic Eddy-Flux Measurement with the Evapotron," *J. Appl. Meteorol.* 4:622 (1965).
16. Dyer, A. J., B. B. Hicks, and K. M. King. "The Fluxatron — A Revised Approach to the Measurement of Eddy Fluxes in the Lower Atmosphere," *J. Appl. Meteorol.* 6:408 (1967).
17. Jones, F. E., and A. Wexler. "A Barium Fluoride Film Hygrometer Element," *J. Geophys. Res.* 65:2087—2095 (1960).
18. Goltz, S. M., C. B. Tanner, G. W. Thurtell, and F. E. Jones. "Evaporation Measurements by an Eddy Correlation Method," *Water Resour. Res.* 6:440—446 (1970).
19. Bean, B. R., and Q. L. Florey. "A Field Study of the Effectiveness of Fatty Alcohol Mixtures as Evaporation Reducing Monomolecular Films," *Water Resour. Res.* 4:206—208 (1968).
20. Brutsaert, W. "A Model for Evaporation as a Molecular Diffusion Process into a Turbulent Atmosphere," *J. Geophys. Res.* 70:5017 (1965).
21. Dankwerts, P. V. "Significance of Liquid-Film Coefficients in Gas Absorption," *Ind. Eng. Chem.* 43:1460 (1951).
22. Dobbins, W. E. "The Nature of the Oxygen Gas Transfer Coefficients in Aeration Systems," in *Biological Treatment of Sewage and Industrial Wastes*, Part 2-1, McCabe, J., and W. W. Eckenfelder, Eds., (New York: Reinhold Publ. Corp., 1956).
23. Shepherd, C. B., C. Hadlock, and R. C. Brewer. "Drying Materials in Trays, Evaporation of Surface Moisture," *Ind. Eng. Chem.* 30:388 (1938).
24. Millar, F. G. "Evaporation from Free Water Surfaces," *Can. Meteorol. Mem.* 1:39 (1937).
25. Himus, G. W. "The Evaporation of Water in Open Pans," *Trans. Inst. Chem. Eng.* 7:166 (1929).
26. Powell, R. W., and E. Griffiths. "The Evaporation of Water from Plane and Cylindrical Surfaces," *Trans. Instr. Chem. Eng.* 13:175 (1942).
27. Gililand, E. R., and T. K. Sherwood. "Diffusion of Vapors into Air Streams," *Ind. Eng. Chem.* 26:516 (1934).
28. Deissler, R. G. "Analysis of Turbulent Heat Transfer, Mass Transfer, and Friction in Smooth Tubes at High Prandtl and Schmidt Numbers," U.S.N.A.C.A. Report No. 1210, 1955.
29. Reginato, R. J., R. D. Jackson, and P. J. Pinter, Jr. "Evapotranspiration Calculated from Remote Multispectral and Ground Station Meteorological Data," *Remote Sens. Environ.* 18:75—80 (1985).
30. Jackson, R. D., P. J. Pinter, Jr., and R. J. Reginato. "Net Radiation Calculated from Remote Multispectral and Ground Station Meteorological Data," *Agric. Forest Meteorol.* 35:153 (1985).

31. Jackson, R. D., M. S. Moran, L. W. Gay, and L. H. Raymond. "Evaluating Evaporation from Field Crops Using Airborne Radiometry and Ground-Based Meteorological Data," *Irrig. Sci.* 8:81—90 (1987).
32. Miller, W., and E. Millis. "Estimating Evaporation from Utah's Great Salt Lake Using Thermal Infrared Satellite Imagery," *AWRA Water Resources* 25:541—550 (1989).
33. Chattree, M., and S. Sengupta. "Heat Transfer and Evaporation from Heated Water Bodies," *J. Heat Transfer* 107:779—787 (1985).
34. Adams, E. E., D. L. Cosler, and K. R. Helfrich. "Evaporation from Heated Water Bodies: Predicting Combined Forced Plus Free Convection," *Water Resour. Res.* 26:425—435 (1990).

CHAPTER 10

Evaporation Pans

INTRODUCTION

A Bulletin of the California Department of Water Resources includes a very useful introduction to pan evaporation measurements.[1] Some of this material will be included in this section.

> Pan evaporation measurements are useful in water use projections, such as forecasts of irrigation water demand for crops, estimates of losses from percolation ponds used for ground water recharge, and the design of ponds for concentrating brines and effluents from waste disposal facilities. Data from evaporation pans indicate the atmospheric evaporative demand of an area and can be used to estimate (1) the rates of evaporation from ponds, lakes, and reservoirs, and (2) evapotranspiration rates (derived by combining rates of soil evaporation and plant transpiration rates).

The Bulletin provided mean monthly total evaporation data from various types of evaporation pans at 478 locations throughout California. The measurements represented the actual depth of water lost to the atmosphere, sometimes called gross evaporation. Of the 478 records, 261 were derived from U.S. Weather Bureau Class-A pans; 10 were from Bureau of Plant Industry pans; 14 were from square floating pans, sometimes called the U.S. Geological Survey (U.S.G.S.) pan; 33 were from land pans of the same size as the U.S.G.S. pan, sometimes called the Colorado pan; and 53 were from Young pans.

The Class-A pan represented the first large-scale attempt to unify evaporation measurements. The appropriate siting for the pan was determined, by the California Department of Water Resources, to be in a large, well-irrigated pasture. At the beginning of each observation of the Class-A pan, the depth of water in the pan was 203 mm (8 in.). Water in the pan warmed rapidly during the morning because it received the full effect of sun and wind. It also cooled rapidly after sundown. During the day it had a higher rate of evaporation than any other pan in common use. It is generally considered the standard for evaporation measurements. It is relatively free from drifting sand or rolling weeds, however, it is not easy to keep the water clean. At certain temperatures growths of algae appear.

The major factors that affect the rate of pan evaporation are:

1. Type of pan
2. Type of pan environment
3. Method of operating the pan
4. Exchange of heat between pan and ground
5. Solar radiation
6. Air temperature
7. Wind
8. Temperature of the water surface

Pan evaporation measurements normally exceed lake surface evaporation, some pans are equipped with screens so that evaporation rates will correspond more nearly to lake surface evaporation. Pan evaporation rates are also affected by the relative clarity of the water and by relative aridity or moistness of the pan's upwind fetch.

A source of inconsistency in evaporation records is the height at which the water level is maintained in the pan. Environmental factors can cause as much as 77% overindication in an arid environment as compared to a well-irrigated environment in the same climatic zone.

Lake evaporation is exceedingly difficult to measure directly. Pan evaporation rates are used to estimate lake *evaporation* by reducing the pan rates by a fraction called the pan coefficient. Class-A pans operated in large irrigated pastures are useful for estimating *evapotranspiration* from crops and native vegetation. Pan data have been used to estimate daily and weekly crop evapotranspiration for scheduling irrigations to maximize crop yield and minimize energy consumption. Pan-to-crop ratios, similar in concept to pan coefficients, are used for adjusting pan evaporation rates to equivalent crop evapotranspiration.

The annual mean pan coefficients for Class-A pans at Lakes Elsinore and Fullerton in California were 0.79 and 0.76, respectively. The corresponding values for Young pans at the same locations were 0.98 and 0.96, respectively. The annual mean pan-to-crop ratios for alfalfa, orchard, and pasture crops were 0.76, 0.83, and 0.76, respectively.

In 1982, the U.S. National Weather Service (NWS) Office of Hydrology published Evaporation Atlas for the Contiguous 48 States.[2] In the present context we will be interested in pan evaporation and particularly the Class-A pan as discussed in this report. Pan evaporation, as used in the report, means "evaporation observed at a standard NWS Class-A pan installation by observers following standard techniques." Free water surface (FWS) evaporation was defined as "evaporation from a thin film of water having no appreciable heat storage." It is considered to be "approximately equivalent to potential evaporation from a shallow water surface and to potential evapotranspiration from a vegetative surface with an unlimited supply of water." FWS can be computed from meteorological factors.

The pan coefficient is defined in the Atlas as the ratio of FWS evaporation to observed pan evaporation. The pan coefficient commonly used to compute FWS evaporation from Class-A pan measurements is 0.7. The pan coefficients in the U.S. varied from 0.64 to 0.88 for the May through October period. When the water in the pan is warmer than the air, the coefficient is greater than 0.7, and vice versa; the tendency for most locations is for coefficients in winter to be lower than those for summer months.

The primary reason for variations in pan-to-lake coefficients is the energy exchange through the sides and bottom of the Class-A pan.

The Class-A Pan

Class-A evaporation pans[1-4] are used by the NWS at hundreds of locations to measure evaporation. They are used in many other countries as well. The data collected from these pans are used to estimate lake evaporation or evapotranspiration rates, although the construction of the pan does not simulate natural conditions.

The pan is cylindrical and constructed of unpainted monel or galvanized iron. The inside diameter is 47.5 in. (121 cm) and the depth is 10 in. (25.4 cm). It is mounted on a wooden platform above the surrounding soil, allowing heat transfer from the air through the bottom and side of the pan. The pan is thus provided with a heat source (or sink) not available to the water in a lake or to soil. To anchor the platform, it is sometimes filled with new soil to within 2 in. (5 cm) from the bottom, allowing some ventilation.

At the beginning of each observation of evaporation rate, the depth of water in the pan is 8 in. (20 cm); the water surface is thus 2 in. (5 cm) below the rim. The water level is usually observed using a fixed-point gauge, hook guage (a pointed shaft extending vertically from the bottom of the pan), surrounded by a stilling well. The position of the tip of the shaft is fixed so that when the water surface just meets the point, the surface is 2 in. (5 cm) below the rim of the pan. Measured amounts of water are added or removed (in the case of rain) to maintain the water surface 2 in. (5 cm) below the rim.

In a standard installation, the water temperature is measured using a thermometer (preferably max-min). A 3-cup anemometer measures the wind speed at a height about 2 ft (61 cm) above the ground. A rain guage is usually placed nearby in an attempt to determine the amount of rain entering the pan.

The Young Screened Pan

The Young screened pan[1,5] (first used by Arthur A. Young, a research engineer of the Division of Irrigation, U.S. Department of Agriculture) is 2 ft (61 cm) in diameter, 3 ft (91 cm) deep, and sunk 33 in. (84 cm) in the ground in operation. It is fitted with a 0.25-in (0.64-cm) mesh screen over an open top. Young was attempting to produce a pan evaporation loss to lake evaporation loss ratio of unity, which the screened pan approximates in the coastal area of Southern California. The mean evaporation rates for Young pans and Class-A pans were compared for 14 California stations. The mean ratio of the Young to Class A rates was 0.75 for inland stations and 0.93 for coastal stations.

The Colorado Sunken Pan

The Colorado sunken pan[4] has a square water surface with 3-ft (91-cm) sides and a usual depth of 1.5 ft (46 cm). It is installed in the ground with its rim approximately 4 in. (10 cm) above the ground surface, so that the water surface is maintained at approximate ground level.

The Sunken Pan of the Bureau of Plant Industry (BPI)

The sunken pan of the BPI[1] is 6 ft (183 cm) in diameter and 2 ft (61 cm) deep. It is buried 20 in. (51 cm) in the ground. It is filled with water up to ground level, 4 in. (10 cm) below the rim. It is usually constructed of the same type sheet metal as the Class-A pan.

The GGI-3000 Pan

The GGI-3000 pan[4] is cylindrical with a conical base, with a diameter of 24.3 in. (61.7 cm), a depth of 23.6 in. (60 cm) at the wall, and a depth of 27.0 in. (68.6 cm) at the center. It is buried in the ground with the rim about 3.0 in. (7.6 cm) above the ground level. It has a surface area of 3000 cm^2, hence the name. It is made of galvanized iron and was developed in the U.S.S.R. It is widely used as a standard, especially in Eastern Europe.

The 20-m^2 Basin Pan

The 20-m^2 Basin pan[4] is cylindrical with a flat base made of 1.0- to 1.3-in. (2.5- to 3.3-cm) boiler plate sheets or concrete. The diameter is 197 in. (5 m) and the depth is 79 in. (2 m). It has a water surface area of 215 ft^2 (20 m^2, hence the name) and is placed in the ground with the rim 3.0 in. (7.6 cm) above ground level. The water level is maintained at approximate ground level. The pan was developed in the U.S.S.R.

Los Angeles County Flood Control District Pans G and L

Pan G[1] is 2 ft (61 cm) in diameter, 3 ft (91 cm) deep, and set in the ground with 3 in. (7.6 cm) of the rim exposed. The water level, which normally is at the level of the ground surface, is indexed by a brass rod pointed at the upper end and set in a concrete block on the bottom of the pan.

Pan L[1] is square with a side length of 3 ft (91 cm), is 18 in. (46 cm) deep and is set 14 in. (36 cm) into the ground.

Floating Pans

A variety of floating pans[1] have been in use. The exchange of heat between the pan and surrounding water, characteristic of floating pans is advantageous for simulation. However, the action of waves in moving water into and out of the pan is a major problem. Also, measurement or determination of water level in the pan is inconvenient.

HEAT BALANCE OF THE CLASS-A EVAPORATION PAN

Since the construction of the Class-A pan does not simulate natural evaporation conditions and since the mounting of the pan on the platform permits heat transfer from the air through the bottom and side of the pan, Riley[3] studied the feasibility of calculating the heat transfer through the walls of the pan from energy balance considerations and computed the order of magnitude of this transfer. He made evaporation measurements using a standard Class-A pan and an insulated Class-A pan (wrapped with a 2-in., 5-cm, layer of polyethylene foam; Perlite was poured into the wooden stand until full, providing at least a 4-in., 10-cm, layer of insulation below the pan).

The heat balance of the standard pan was given as

$$G = R + H + S + U + LE \tag{1}$$

where G = heat added to the water (calculated from measurements of water temperature)
R = net radiation (measured using a miniature net radiometer mounted 6.4 in., 16 cm, above the water surface in the center of the pan)
H = sensible heat transfer to the air (estimated from the corresponding term in an insulated Class-A pan)
S = heat transfer through the side of the pan
U = heat transfer through the bottom of the pan
LE = latent heat transfer calculated from pan evaporation measurements

The corresponding equation for the insulated pan, with insulated pan data denoted by the subscript I, is

$$G_I = R_I + H_I + LE_I \tag{2}$$

G_I and LE_I were calculated directly as they were for the standard pan, R_I was estimated from R by correcting for the difference in the water temperature in the standard and insulated pans. H_I was calculated as the remainder term.

The evaporation rate for the standard pan was greater than the evaporation rate for the insulated pan in each 6-hr period throughout the day, beginning at midnight, with maximum difference occurring in the evening. The average

Figure 1. U.S. Weather Bureau Class-A pan. (Reprinted from Department of Water Resources, State of California by permission.)

daytime water temperature of the standard pan was 2°C above that of the insulated pan as a result of heat transferred through the walls.

The heat transfer through the walls of the standard Class-A evaporation pan made a significant contribution to the heat balance and, consequently to the rate of evaporation. It appeared that the increased nighttime evaporation rates resulted from the heat supplied through the walls in daytime.

The large heat transfer through the walls of the Class-A pan needs, according to Riley, to be taken into account, especially when estimating lake evaporation from pan data. Also, the turbulent transfer of water vapor is enhanced by the elevation of the pan above ground level. A Class-A pan is shown in Figure 1.

SIMULTANEOUS RECORDING OF PAN EVAPORATION AND RAINFALL

Bloemen[6] described an instrument devised to simultaneously record pan evaporation and rainfall with high accuracy. It was very clear to Bloemen that very accurate short-period measurements of pan evaporation were beyond the capabilities of a micrometer gauge. Also, it had been reported[7] that the rate of pan evaporation is influenced by a change in water level. Consequently the pan-evaporimeter described was operated on the constant-water-level principle.

To obtain the maximum advantage from the constant-water-level principle, Bloemen set forth the following conditions to be met: "(1) both the supply to and discharge from the pan should be recorded; (2) the dimensions of the reservoirs should be such that small changes in water content of the pan give large changes in water level in the reservoirs; (3) the rate of supply to or discharge from the pan has to follow the rates of loss by evaporation or gain by precipitation as closely as possible."

The apparatus consisted of a shallow evaporation pan communicating with a float tank which maintained the water in the pan at a constant level by addition or extraction of water in accordance with evaporation or excess rainfall, respectively. Evaporated water was supplied from a reservoir, the level of which was recorded on a chart by means of a float. Discharged water was caught in a receptacle, the water level in which was recorded also. Evaporation was recorded upwards on the chart, rainfall downwards.

The evaporation pan had a diameter of 24.6 in. (62.5 cm) and a depth of 1.6 in. (4 cm). The water level was maintained at 0.4 in. (1 cm) below the pan rim.

Observing that evaporation from a deep pan lags behind that of a shallow one (evaporation continuing into the night at a higher rate due to heat storage during the day) and that during rainfall when incoming radiation is low the evaporation increases due to the fast transfer of heat (and consequent increase in water vapor pressure) stored in the pan to the water surface cooled by rain, Bloeman used a 1.6-in. (4-cm) pan shielded from precipitation.

In operation of the pan evaporimeter, water was supplied to the pan after 0.001 in. (0.03 mm) had evaporated from it. This, then, is a measure of the sensitivity of the device. Shielding the pan from the rain with a plastic transparent dome, 43.3 in. (110 cm) in diameter, set 11.8 in. (30 cm) above the pan gave a close correlation with calculated crop evaporation. A ground-level rain gauge of the same diameter as the evaporation pan was used and evaporation and rainfall were recorded simultaneously. This combination was in use for routine measurements at various sites in the Netherlands.

DIFFERENCES IN EVAPORATION PAN DATA

Gunderson,[8] in the Everglades National Park in south Florida, U.S., set up four experiments to document and account for differences in evaporation data calculated using evaporation pans equipped with float-activated recorders and pans with hook gauge/rain gauge instrumentation. Conventionally, evaporation data were collected using different techniques at three sites. At two sites, standard NWS Class-A pans were used; evaporation was determined by combining the daily hook gauge reading of pan water levels with rainfall recorded in adjacent Belfort-type rain gauges. At the third site, a float-activated recorder tracked water level changes in the pan; evaporation was calculated by summing all decreases in water level.

Differences in evaporation were noted among the sites in June 1985. Observed differences were surmised to be due to site variation or to the method of calculating evaporation. On days with rain, evaporation was estimated by measurements of precipitation and water levels. On days without rain, the estimation was based solely upon the difference in measured water levels. Discrepancies could have been a result of rainfall measurements or of the method of measuring the water level. The four experiments were conducted to determine reasons for the observed differences.

The first experiment examined whether differences in evaporation could be attributed to the method of calculation within the pan, thus eliminating site-to-site variation. A standard Class-A pan was equipped with both a hook gauge and a float-activated recorder. The recorder was used to measure water levels in the pan, and thus was used to monitor both increases due to rainfall and decreases due to evaporation. A standard NWS rain gauge adjacent to the pan was used to independently estimate rainfall inputs to the pan. The pan and the rain gauge were mounted on a wooden platform.

The second experiment was used to detect differences, if they existed, between the methods of measuring water levels, by comparing readings from a hook gauge with readings from the float-activated recorder. Both devices were placed within a level water-filled standard Class-A pan.

The third experiment tested for float-lag in the recorder since it had been suggested by Miereau (personal communication) that float lag accounted for the observed discrepancies in evaporation amounts. Changes in the water level in the pan would be underestimated by a slow or incomplete response of the float system. To test the float-lag assertion, a float-activated recorder was set up within a bucket placed indoors. Water was withdrawn from the bucket at slow rates; then, after 20 min of withdrawing water, 3.1 to 6.1 $in.^3$ (50 to 100 ml) of water was added to the bucket. The float lag was then evaluated using pen tracings on the recorder chart. If float lag were significant, the pen tracings would be rounded at deflection or change points; sharp deflection points would indicate little or no lag.

Gunderson[8] suspected that the differences in rain capture between the rain gauge and the pan might account for the differences in calculated evaporation, since Huff,[8a] Hamilton,[9] and Green[10] had indicated that smaller-diameter gauges catch more rain than larger-diameter gauges because of greater edge to area ratio. He set up the fourth experiment to determine whether differences existed between the amount of rain captured by an 8-in. (20-cm) diameter rain gauge and a standard Class-A pan. Two standard Class-A pans, three NWS rain gauges, orifice of 8 in. (20 cm), and three Belfort-type gauges, orifice of 8 in. (20 cm), were arranged randomly within an 18-ft (5.5-m) by 18-ft (5.5-m) area. Rain events during a week in August of 1985 were sampled. The water levels in the pans were measured 20 min before and after rain events, using hook gauges; recorders were used also to measure the rain-caused water level rise in each pan. The rain in the rain gauges was measured for each event using a graduated dipstick. Data from the Belfort-type gauges were recorded on charts.

A two-way analysis of variance was used to determine whether a significant difference in evaporation calculations could be attributed to the method of calculation. The response variable was daily evaporation and the two factors were the water level measuring device (hook gauge and pan recorder) and the rain gauge type (rain gauge and pan recorder). The mean evaporation was calculated by the four methods (water level/rain gauge). The mean evaporation calculated with the rain gauges was significantly higher than the mean calculated using the within-pan rain estimates. The difference between means calculated with the two water level measuring devices was not statistically significant.

The paired readings of water levels within the pan were compared using a paired t-test with the test statistic being the difference between the two readings and using the null hypothesis that this difference is equal to zero. The mean difference between the gauge types was statistically significant. As in the first experiment, the readings with the hook gauge were higher than those from the pan recorder. The pen tracings for the float-lags tests indicated little evidence of float lag.

The data from four rain events indicated that the rainfall was sampled differently by different gauge types. The rainfall registered in the pan was smaller over all events. The data were analyzed using a one-way analysis of variance. The response variable was the amount of rain, the factor tested was type of gauge. The analysis indicated that the amount of rainfall recorded can differ significantly depending on the type of gauge. In all the rainfall events, no significant differences could be ascertained between groups of Belfort and Weather Service gauges. The two methods for measuring rainfall in the evaporation pans were found to be statistically similar.

The differences in the calculated evaporation can be attributed to differences in measured rainfall between the standard rain gauge (8 in., 20 cm) and the evaporation pan. The evaporation pan appeared to catch less rainfall than the other gauges used. The author concluded that consistent instrumentation should be utilized in networks in order to assure consistent results; and that if a network of evaporation pans were to be used for monitoring, modeling, or other purposes, then the same devices for measuring water level and rainfall should be used within all pans to insure consistency of data collection.

EVAPORATION FROM NONMARINE BRINES

Turk[11] studied evaporation from nonmarine brines on the Bonneville Salt Flats, UT. The study was undertaken to establish relative evaporation rates, under natural field conditions, among brines containing various amounts of dissolved solids. The results were to be used to determine the optimum size of evaporation ponds (for recovery of potash by solar evaporation in the Bonneville case) and to help in predicting the limits of evaporation from the more concentrated brines. Also the data might be useful in studies of the disposal of waste brines from desalination processes and oil field brines.

There were eight circular evaporation pans, 30 in. (76 cm) in diameter and 10 in. (25 cm) deep, installed on the bottom of an evaporating pond, simulating actual pond conditions as closely as possible. A ninth pan, a Class A pan, was placed on the salt crust outside the pond at Bonneville Limited's weather station. Fluid levels were monitored using a needle guage, 7.25 in. (18.4 cm) high installed in a stilling well; evaporation losses were determined by measuring the volume of water needed to refill the pan to the 7.25 in. (18.4-cm) level.

Specific gravities of the solutions were measured with a set of hydrometers to the nearest 0.001. Temperatures were measured using a standard therometer.

Pan Number 1 and the Class-A pan were filled with tap water (specific gravity about 1.00) to serve as control pans; pan Numbers 2 through 8 were filled with brines from various parts of the pond system. Small amounts of green dye were added to the heavier brines in an attempt to speed evaporation; the amounts were not large enough to adversely affect the experiment. As an indication of the approximate range of characteristics of the brines, the following are for pans Numbers 2 and 8 (the data for pan Number 2 is the first number): specific gravity, 1.20 and 1.34; percent by weight of KCl, 4.20 and 0.22; percent by weight of $MgCl_2$, 5.13 and 32.20; percent by weight of NaCl, 15.75 and 0.60; and percent by weight of SO_4, 0.94 and 2.25. Some of the brines were mixed or diluted with fresh water to create a fluid of the desired specific gravity. After the desired specific gravity was attained, water lost by evaporation was replaced by tap water thus keeping the specific gravity of the brine near the original value.

Measurements were typically made at 11 a.m., the temperature in each pan and in the surrounding pond was measured and the specific gravities of the seven brine samples were determined. Each pan was then refilled to the top of the needle gauge, and the volume of water added was recorded. Specific gravities were then measured to insure that they were close to the original values.

The volume of water evaporated (added) was converted to inches of evaporation to determine the mean daily evaporation for each time interval. If there were precipitation during the time period, the amount of rainfall measured by three rain gauges approximately equidistant from the pans was added to the measured evaporation. Rainfall on the pans was assumed to be the mean of the rainfall at the three rain gauges.

Evaporation rates were found to decline as specific gravity increased, that is, as the concentration of dissolved solids increased. One could estimate the rate of evaporation from brine of any specific gravity by monitoring the rate of evaporation from a fresh water pan.

A comparison of fresh water evaporation from the Class-A pan and from pan Number 1 showed that the pan in the pond averaged only 41% as much evaporation as the pan on the salt surface. Thus, the pan coefficient for converting fresh water evaporation from the usual top of the ground pan to fresh water evaporation under pond conditions at Bonneville is 0.41; however, this value is valid only for evaporation of fresh water, and only during the summer.

Turk concluded that: (1) the specific gravity of a brine affects its evaporation rate in a predictable fashion, at least in summer months; therefore, evaporation from brines of various specific gravities can be estimated rather accurately by monitoring a single fresh water evaporation pan placed in a pond; (2) condensation, or "negative evaporation," may occur during winter (and rarely during summer), especially on heavier brines; (3) a thin salt crust on the brine surface may retard evaporation from the heavier brines; and (4) physical conditions near the brine surface of a pond differ greatly from those surrounding a pan outside the pond.

EFFECT OF SALINITY AND IONIC COMPOSITION ON EVAPORATION

Salhotra et al.[12] discussed the use of pan evaporation data to evaluate the effect of salinity and water chemistry on evaporation, based on data collected near the Dead Sea in Israel. They expressed the rate of evaporation per unit area of fresh water surface as

$$E_{fresh} = f(W)\{e_{sat}(T_s) - \psi e_{sat}(T_a)\} \tag{3}$$

where E_{fresh} = evaporation rate

$e_{sat}(T)$ = saturation vapor pressure over a fresh water surface as a function of temperature, T

T_s and T_a = temperatures at the water surface and in the air, respectively
ϕ = relative humidity expressed as a fraction
f(W) = empirical function of wind speed, W

The reduced evaporation from a saline surface was expressed as

$$E_{sal} = f(W)\{e_{sat}'(T_s,S_s) - \psi e_{sat}(T_a)\} \quad (4)$$

where E_{sal} = evaporation rate from the saline water body
$e_{sat}'(T_s,S_s)$ = saturation vapor pressure above a saline surface, that decreases with surface salinity, S_s
f(w) = same in both equations

The relationship between saturation vapor pressure and salinity depends on the particular salt in solution. The activity coefficient, β, represents the ratio of vapor pressure over salt walter of salinity S to the vapor pressure over fresh water at the same temperature. As an approximation,

$$e_{sat}'(T,S) = \beta(S)e_{sat}(T) \quad (5)$$

or

$$E_{sal} = f(W)\{\beta(S_s)e_{sat}(T_s) - \psi e_{sat}(T_a)\} \quad (6)$$

The approach of accounting for the effect of salinity on evaporation can be contrasted with the common approach based on an empirical ratio of salt water to fresh water evaporation assuming similar meteorological conditions embodied in

$$E_{sal} = \alpha f(W)\{e_{sat}(T_s) - \psi e_{sat}(T_a)\} \quad (7)$$

$$\alpha = E_{sal}/E_{fresh} \quad (8)$$

represents the reduction in evaporation due to salinity.

In the few pan evaporation experiments using saline water in the literature, results were presented as the ratio of salt water evaporation to fresh water evaporation. Salhorta et al.[12] undertook to study the method based on the effect of salinity on saturation vapor pressure.

The experimental setup consisted of eight cylindrical evaporation pans, 9.8 ft (3 m) in diameter and about 3.3 ft (1 m) deep, sunk in the ground at the southern end of the Dead Sea. The pans were typically filled to a depth of about 3.1 ft (0.94 m) with mixtures of Mediterranean and Dead Sea water in various proportions. Air temperature, relative humidity, and temperature at the water surface and at various depths within the pans was measured. The pans, identical in shape and size and exposed to the same meteorological conditions, were used to study the effect of differences in salinity and chemical composition on the rate of evaporation.

Fresh makeup water was added to each pan, at intervals ranging from three days to several weeks, to compensate for the evaporative water loss and to maintain the same chemical composition. At each addition the contents of the pan were mixed vigorously for 2 to 3 hr to redissolve any precipitated salts. Before and after each addition, water surface elevations were measured and chemical assays were taken for each pan.

Data were collected for 36 cycles, a cycle consisting of the interval between two makeup-water additions. For each cycle, means of water surface temperature, temperature at 3.9-in. (9.9-cm) depth, air temperature, humidity, duration of the cycle, makeup water added, and net evaporation were recorded; densities and detailed analysis for each cycle were reported. The accuracy of the net evaporation rates was estimated to be 0.001 in./day (0.03 mm/day), the accuracy of water temperature measurements was estimated to be 0.2°C. A maximum temperature difference between the pans of greater than 5°C occurred during the month of July. Unfortunately, none of the pans contained fresh water.

The effect of salinity, temperature, and ionic composition on saturation vapor pressure was determined from data from the 23 cycles with complete meteorological measurements. For a given cycle, differences in evaporation rates among pans reflected effects of salinity and ionic composition. Rates of evaporation computed based on the α approach had a larger uncertainty than those computed using the β approach. Also, the cycle-to-cycle variation for α was much larger than for β.

Various analytical methods to compute the effect of salinity on saturation vapor pressure based on ionic composition of the solutions were described and applied to the Dead Sea data; these approaches could be applied in many engineering applications. Comparison of the average decrease for the diluted Dead Sea brine ($\beta = 0.83$) with that for the concentrated seawater of the same salinity ($\beta = 0.87$) illustrates the dependence of evaporation on chemical composition.

PAN-EVAPORATION DATA AS A PRACTICAL TOOL FOR ESTIMATING POTENTIAL EVAPOTRANSPIRATION

Pruitt[13] cited studies which indicated that pan-evaporation data could be useful in estimating potential evapotranspiration, beginning with the classic studies of Briggs and Shantz[14,15] published in 1916 and 1917. The Briggs and Shantz studies revealed a striking correlation between evaporation of water from a very shallow pan and diurnal patterns of transpiration. For our purposes here we use the definition of potential evapotranspiration of Thornthwaite:[16]

the maximum amount of water that could be lost from large, well-irrigated fields by evaporation from the soil and transpiration from the plants. Decker[17] gives the definition, "amount of water transferred from a wet surface to vapor in the atmosphere", representing "the greatest possible mass transfer and energy dissipation from the earth's surface by vaporization of water."

Among the studies cited by Pruitt: McIlroy and Angus[18] reported results using highly reliable, replicated weighing lysimeters planted to a pasture mixture, noting an extremely close relationship for monthly evapotranspiration and evaporation over a three-year period at Aspendale, Australia. The following quotation from their work is quite interesting: "The excellent correlation here is more than a mere reflection of the basic dependence of heat fluxes on available energy. It demonstrates that all the factors contributing to the evaporation, including the proportioning of the total atmospheric energy flux between sensible and latent heat, are integrated over a period in an essentially similar way by the two systems, freely transpiring crop and free water surfaces." Their results, however, illustrated that the ratio evapotranspiration (measured by lysimeter)/evaporation (measured by evaporation pan) depends on the type of pan or tank. The average ratio varied from 0.84 for the Class-A pans, to 1.05 for the standard Australian evaporation tank, and 1.20 for the 1.6-m (5.2-ft) weighing evaporimeter. The ratio 0.84 for the Class-A pan was in good agreement with Davis, CA data.[19,20]

The value of ET/E has been shown to vary with the particular vegetation. The ratio of near 1 has been observed for alfalfa; 0.7 to 0.85 for grass; 0.8 for corn, grapes, and peaches with no cover crop; 1.05 for Delicious apples with a grass cover crop; 0.9 for sugar beets, soybeans, red beans, late potatoes, and wheat; around 1.0 for green peas, early potatoes, raspberries, and peaches with an alfalfa cover crop; and an average of 0.95 for seven years for alfalfa in arid central Washington state.

A principal emphasis of Pruitt's paper is based on his observation that the large effect that the local upwind environment can have on evaporation had not been considered. He cited the work of Young[21] and his discussion of the problem in relation to lake evaporation that "should have caused those who later considered using pans for predicting evapotranspiration to pay more attention to standardization of local environment of pans." Pruitt[22] had found that evaporation from a 3-ft (0.9-m) pan in a large dry field exceeded the evaporation from a pan of similar size located in an irrigated grass site in a 580-ft (177-m) by 480-ft (146-m) fallow field, all but two of the pans were placed in the middle of a circular area of irrigated Kentucky bluegrass ranging from 12 ft (3.7 m) to 100 ft (30.5 m) in diameter. Other pans were placed about 150 ft (45.7 m) in from the south edge of a 13-acre (5.3×10^4-m^2) irrigated grass plot 1169 ft (356 m) long in the north-south direction and 480 ft (146 m) wide. Average rates of evaporation in inches per day were plotted against the upwind fetch (extent) of cropped land. The effect of the environment produced by immediate upwind conditions was very marked. The decrease in evaporation was nearly linear with the logarithm of the upwind fetch for all of the months of the study.

The decrease in pan evaporation as the fetch of grass increased beyond 4 ft (1.2 m) was attributed by Pruit directly to increased local advection as the humidity increased and the air temperature dropped due to transpiration from the grass. It was obvious that the standardization of the local environment of evaporation pans is highly desirable if they are to be used to estimate evapotranspiration by crops. The proximity of major differences in crop height or roughness should also be a consideration.

Pruitt also observed a serious effect on the ET/E relationship of strong advection, especially in conjunction with strong dry winds.

Pruitt concluded that: ''Where reliable measurements have been made and where pan evaporation has been fairly well standardized throughout growing seasons, results by and large indicate that pan-evaporation data could be very useful as a practical tool for estimating potential evapotranspiration.''

REFERENCES

1. Goodridge, J. R. ''Evaporation from Water Surfaces in California,'' California Department of Water Resources Bulletin 73—79, 1979.
2. Farnsworth, R. K., E. S. Thompson, and E. L. Peck. ''Evaporation Atlas for the Contiguous 48 United States,'' NOAA Technical Report NWS 33, 1982.
3. Riley, J. J. ''The Heat Balance of Class A Evaporation Pan,'' *Water Resour. Res.* 2:223—226 (1966).
4. Brutsaert, W. *Evaporation into the Atmosphere* (Dodrecht, Holland: D. Reidel Publ. Co., 1982), pp. 251—252.
5. Young, A. A. *Evaporation from Water Surfaces in California, A Summary of Pan Records and Coefficients, 1881 to 1946,* California Division of Water Resources Bulletin No. 54, 1947.
6. Bloeman, G. W. ''A High-Accuracy Recording Pan-Evaporimeter and Some of its Possibilities,'' *J. Hydrol.* 39:159—173 (1978).
7. Boynthon, C. W. ''Evaporation Studies Using Some South Australian Data,'' *Trans. R. Soc. South Aust.* 73:198—219 (1950).
8. Gunderson, L. H. ''Accounting for Discrepancies in Pan Evaporation Calculations,'' *Water Res. Bull.* 25:573—579 (1989).

8a. Huff, F. A. ''Comparison Between Standard and Small Orifice Raingauges.'' *Trans. Am. Geophys. Union* 36: 689—694.

9. Hamilton, E. L. ''Rainfall Sampling on a Rugged Terrain,'' U.S. Department of Agriculture Tech. Bull. 1096, 1954.

10. Green, M. J. "Effect of Exposure on the Catch of Raingauges," *N.Z. J. Hydrol.* 9:55—71 (1970).
11. Turk, L. J. "Evaporation of Brine: A Field Study on the Bonneville Salt Flats, Utah," *Water Resour. Res.* 6:1209—1215 (1970).
12. Salhotra, A. M., E. E. Adams, and D. R. F. Harleman. "Effect of Salinity and Ionic Composition on Evaporation: Analysis of Dead Sea Evaporation Pans," *Water Resour. Res.* 21:1336—1344 (1985).
13. Pruitt, W. O. "Empirical Method of Estimating Evapotranspiration Using Primarily Evaporation Pans," in *Conference Proceedings: Evapotranspiration and its Role in Water Resources Management* (St. Joseph, MI: American Society of Agricultural Engineers, 1966), 57—61.
14. Briggs, L. J., and H. L. Shantz. "Daily Transpiration During the Normal Growth Period and its Correlation with the Weather," *J. Agr. Res.* 7:155—212 (1916).
15. Briggs, L. J., and H. L. Shantz. "Comparison of the Hourly Evaporation Rate of Atmometers and Free Water Surfaces with the Transpiration Rate Medicago Sativa," *J. Agric. Res.* 9:277—293 (1917).
16. Thornthwaite, C. W. "An Approach Toward a Rational Classification of Climate," *Geog. Rev.* 38:55—94 (1948).
17. Decker, W. L. "Potential Evapotranspiration in Humid and Arid Climates," in *Conference Proceedings: Evapotranspiration and its Role in Water Resources Management* (St. Joseph, MI: American Society of Agricultural Engineers, 1966), 23—26.
18. McIlroy, I. C., and D. E. Angus. "The Aspendale Multiple Weighed Lysimeter Installation," C.S.I.R.O. Division of Meteorology and Physics Tech. Paper, 14; 1963.
19. Pruitt, W. O. "Correlation of Climatological Data with Water Requirements of Crops," 1959—1960 Annual Report, Department of Irrigation, University of California, 1960.
20. Pruitt, W. O. "Evapotranspiration—A Guide to Irrigation," *California Turfgrass Culture* 14:27—32 (1964).
21. Young, A. A. "Evaporation from Water Surfaces in California," Division of Water Resources, Department of Public Works, State of California Bull. No. 54, 1947.
22. Pruitt, W. O. "Relation of Consumptive Use of Water to Climate," *Trans. Am. Soc. Agric. Eng.* 3:9—13 (1960).

CHAPTER 11

Lysimetry

INTRODUCTION

A lysimeter is a soil-filled container placed in a field on which vegetation can be grown. If the lysimeter is isolated mechanically from the surrounding soil by a small air space so that it can be weighed, it is called a weighing lysimeter.

Evaporation or evapotranspiration can be directly determined, using a weighing lysimeter, as the loss in weight (mass) of the lysimeter with time. This operation of the weighing lysimeter is rather like weighing an aircraft by weighing the airport before and after takeoff.

Lysimeters have a long history. The earliest actual lysimeter investigation which Kohnke and Dreibelbis[1] were able to study was by De la Hire in Paris in 1688.[2] De la Hire, a mathematician and meteorologist, used leaden vessels of three different depths (20 cm, 8 in.; 41 cm, 16 in.; and 2.4 m, 8 ft) which he filled with sandy loam from the park of the castle of Louis XIV. Among his findings were that lysimeters in grass evaporated more rapidly than in fallow.

Sturtevant[3] built the first lysimeter in the United States on his farm in Massachusetts in 1875. He later, in 1882, installed a set of three lysimeters at the experiment station in Geneva, New York. All of these lysimeters were of the soil-block, monolithic, type (a case was built around the sides of a block of soil found in the field). They were constructed of white oak planks lined with heavy sheet copper and the bottom was made of a heavy flat section of "boiler iron."

LYSIMETRY AND EVAPOTRANSPIRATION

In his paper on the measurement of evapotranspiration, Tanner[4] reviewed various methods for measuring evapotranspiration, including lysimetry. He defined evapotranspiration as consisting of "the conversion to vapor and mixing with the atmosphere of the liquid water at the earth-atmosphere boundary; this may be soil moisture, ponded water, water intercepted on surfaces, and water in plants." A major portion of the water in the hydrologic cycle of most regions is the flux of water vapor in the atmosphere.

Lysimetry is the only method in which each of the quantities in the water balance is completely known. A lysimeter encasing a block of soil in situ is called a "monolith" lysimeter. In a "filled" lysimeter, the soil is returned in the same profile order and packed to the same density as the natural soil. The lysimeter must be sited in identical surrounds and with representative fetch. The lysimeter and surrounds should be planted, fertilized, watered, and otherwise managed in the same manner.

Tanner defined potential evapotranspiration as "that occurring at a 'wet' surface...and is limited by the heat supplied under given micro-meteorological conditions and not by the water supply to the surface." One of the earliest approaches to estimating potential evapotranspiration from soils and vegetables, and evaporation from bodies of water was through correlations to the evaporation measured from pans. Over approximate time periods, there is a high correlation of pan evaporation to potential evapotranspiration even though the pan is a poor analog of a vegetation surface. If the correlation over weekly or monthly periods is relatively high, it may be quite low when water is limiting. The siting of the pan affects the measured pan evaporation greatly, therefore, calibration of an evaporation pan to a given locale is mandatory.

HYDRAULIC LYSIMETERS

Black et al.[5] described the construction and calibration of two 35-metric-ton hydraulic lysimeters with suction drainage systems. The lysimeters were designed to measure evaporation from row crops as well as from other agricultural surfaces. One was circular and the other was rectangular. The tanks were weighed using large hydraulic load cells made from butyl-nylon industrial rubber pipe. The desired resolution for hourly or half-hourly evaporation was about 0.025 mm (0.001 in.) of water.

Both lysimeters consisted of an inner and outer tank. They were installed in sand plains. The tanks were located 30 m (98 ft) apart in a field with a 200-m (656-ft) southwesterly fetch, and a 50-m (164-ft) fetch to the north and west. Instrument wells were 55 m (180 ft) north of the tanks.

The dimensions of the inner tank were 5.5 m (18 ft) by 2.1 m (6.9 ft) by 1.5 m (4.9 ft) for the rectangular lysimeter, and 4 m (13 ft) in diameter and 1.5 m (4.9 ft) deep for the round tank; the inner tank areas were 11.7 m^2 (126 ft^2) and 12.4 m^2 (133 ft^2) for the rectangular and round tanks, respectively. There was a 10-cm (25-in.) gap, fitted with a stainless steel cover, between inner and outer tank walls on both lysimeters. The load cells were inserted under the tanks and connected to copper lines leading to instrument wells. The cells were filled with antifreeze, from which air had been excluded, until the tanks lifted approximately 2.5 cm (1 in.).

The inner tanks rested on four hydraulic load cells. As the inner tanks were filled with soil, germanium diode thermometers were positioned. When filling was completed, the inner tank lips were about 6 mm (0.2 in.) below those of the outer tanks. The pressures in the load cells varied with the weight (mass) of the lysimeter. The changes in lysimeter weight were expressed as equivalent depths of water in mm over the surface area of the inner tank.

Evaporation of 0.025 mm (0.001 in.) produced about 0.01 mm of mercury (0.0002 psi) pressure change in the load cells which were at a total pressure of 1000 mm of mercury (19 psi). The temperature coefficient of the lysimeter consisted of the temperature coefficient of the pressure transducer system and the temperature coefficient of the butyl-nylon load cells. The latter is the major disadvantage of the hydraulic load-cell lysimeters when compared to mechanical balance lysimeters.

The sensitivities of the weighing systems were determined by adding or subtracting weights on the lysimeter; the sensitivities were 0.490 mm mercury/mm water for the round lysimeter and 0.254 mm mercury/mm water for the rectangular lysimeter. The resolution of the weighing systems was 0.02 mm (0.0008 in.) of evaporation. Over the range of 50 mm (2 in.) equivalent depth of water, both weighing systems were linear, and there was no measurable hysteresis due to either the load cells or the covers. Weights positioned over the center of each tank gave the same results as weights placed over the four load cells. The water [0.02 mm (0.0008 in.)] removed by the drainage system agreed with the weighing system within the resolution of the systems.

The steel walls of the lysimeter, because of their high heat conductivities, caused the lysimeter thermal regime to be different from that in the surrounding soil. Therefore, the lysimeter evaporation was not representative of the evaporation from the surrounding field surface. Lysimeter evaporation measurements would, consequently, be low since energy, which otherwise would have contributed to evaporation, would have gone into soil heat flux. The reverse would be true in the afternoon. The authors noted that the thermal regime could be made more representative of that in the surrounding field by placing an insulating layer against the inside of the inner tank walls or by using thin plastic or stainless steel walls. The thermal errors were expected to be much smaller with a crop in the lysimeter.

Evaporation measurements were made using the lysimeters and two micrometeorological methods. The sensible heat flux, H, was measured using two independent methods: the eddy correlation method and the aerodynamic method. The daily values of evaporation obtained by summing the hourly values of each method agreed within 5%. The authors considered the performance of the lysimeters to appear to be adequate under the severe conditions of this test. Again, they anticipated less thermal error when the lysimeters were used under vegetation.

LARGE WEIGHING MONOLITHIC LYSIMETERS

Marek et al.[6] described four large weighing monolithic lysimeters designed, constructed, and installed at Bushland, Texas. Each has a surface area of 9 m^2 (97 ft^2) and a soil depth of 2.3 m (7.5 ft). Each soil monolith, of Pullman clay loam, has a mass of 45 Mg including the mass of the container. Each

Figure 1. A weighing lysimeter with associated micro-meteorological instrumentation. (Reprinted from Conservation and Production Laboratory, U.S. Department of Agriculture—Agricultural Research Service, Bushland, Texas.)

lysimeter was installed in a 20-ha (49-acre) field with one lysimeter located in the center of each quarter of the field.

The lysimeters were designed to be used in research programs requiring precise measurements of evapotranspiration for short-time periods (less than 1 hr), accurate measurements of water interception from both rainfall and sprinkler applications, and accurate measurements of evapotranspiration under water deficit conditions. Weighing lysimeters were considered to be the most versatile research equipment for making these measurements. Design requirements for the lysimeters included a mass resolution equivalent to 0.05 mm of evapotranspiration, a surface dimension of 3 m to permit row spacings from 0.2 to 1.0 m, a soil profile depth of 2.0 m, a soil monolith design to preserve natural field structure, and a narrow air gap to preserve the thermal regime of the soil profile. Preliminary results indicated that the initial requirements established for the design appeared to have been met. A weighing lysimeter is shown in Figure 1.

ACCURACY OF LYSIMETERS

There are, as in all measurement systems, limitations on the accuracy with which lysimeters measure water loss from the soil volume. The overall accuracy of most lysimeters is considered excellent, a value of 0.1 mm is attainable on a daily basis.[7]

Harrold and Dreibelbis[8] reported a precision of 0.25 mm. Pruitt and Angus[9] reported an accuracy of 0.03 mm on a large weighing lysimeter. van Bavel and Myers[10] developed their lysimeter systems to have an accuracy of 0.025 mm. Dugas et al.[11] reported a sensitivity and resolution of 0.02 mm. The facilities of Howell et al.[12] had a resolution of between 0.02 and 0.05 mm.

Minimizable factors that affect the accuracy of lysimeters are wind and the ratio of the surface area to the depth of the lysimeter.

Accuracy of a Hydraulic Lysimeter

Using a prototype size hydraulic weighing lysimeter, Kruse and Neale[13] made a laboratory study of accuracy and error caused by variable pillow contact area, temperature changes in various components of the indicating mechanism, and rate of loading of the lysimeter.

The lysimeter used in the study consisted of two inner tanks, 0.86 m (2.8 ft) × 1.52 m (4.99 ft) × 1.22 m (4.0 ft) deep with an outer tank 1.57 m (5.15 ft) × 1.83 m (6.0 ft) × 1.30 m (4.27 ft) deep. Each inner tank rested on two pillows filled with a solution of equal parts of water and ethylene glycol antifreeze. In an adjacent compartment, a "dummy" pillow with a hydraulic indicating system similar to that of the two "active" tanks was loaded with constant force. The compartment was connected to the outer tank so that similar temperatures were maintained around all pillows.

For ease of handling in the laboratory, water instead of soil was used to load the inner tanks. The total mass of each inner tank, 1877 kg (4138 lb), was approximately two thirds of the expected mass of a soil-filled tank in the field.

The authors concluded that variation in calibration coefficients (expressed as millimeters change in manometer reading for each millimeter of equivalent water depth added or removed from the lysimeter tank) was the result of random error, not the method of loading the lysimeter, and that the lysimeter weighing mechanism gave equally good indications for slow or rapid loading. This represents a variation of −7.5% and +7.0% about the mean value of 4.83 mm/mm. Variations in the coefficients resulted from variations in the corresponding manometer readings.

Hysteresis was not evident in the laboratory tests and neither was mechanical friction between moving parts. Accuracy of lysimeter measurements would be affected if the manometer temperature changed while the lysimeter load was changing. Manometer readings tended to change linearly with changes in pillow temperature. If one assumes that the error in manometer readings in the field is about the same as in the laboratory, one can estimate errors in the determination of mass changes in the field. If the assumption is valid, the expected error in the field should not exceed about ±4% for single determinations with small temperature changes in system components at the beginning and end of the period of measurement.

The authors conclude finally: "Careful and consistent procedures must be observed when hydraulic weighing lysimeters are used. Under such conditions, these lysimeters can be practical research tools."

EVAPORATION RATE AND LYSIMETERS

Kimball and Jackson[14] observed that, after irrigation, the meteorological variables that characterize the different seasons of the year produce large changes in the rate of evaporation from exposed surfaces. Bare soil dries rapidly at the surface, and the evaporation rate from it depends upon the meteorological variables and the length of time since the last irrigation or rain. On the other hand, the time elapsed since the last addition of water did not materially affect the seasonal evapotranspiration from crops in Arizona or the seasonal evaporation rates from open water surfaces (provided that the crop had a full canopy and that the water is not withheld too long). Kimball and Jackson presented data that illustrated how the evaporation rate from bare Adelanto loam soil in Phoenix, Arizona changed with the season and the time since the last irrigation.

Precision weighing lysimeters installed in a field were used to measure water loss. The soil in the lysimeters and in the field was Adelanto loam. Eleven times during ten years, the lysimeters and the surrounding field were irrigated while the soil was bare. The subsequent loss of water from the lysimeters was followed for one week or longer. The irrigations and subsequent drying cycles were observed during every month of the year. The assembled soil drying data revealed seasonal effects on the drying of Adelanto soil.

The lysimeter measurements of evaporation showed that the evaporation rate on the first day of drying after irrigation was about 9 mm/day (0.35 in./day) in summer and 2 mm/day (0.08 in./day) in winter. By the seventh day of drying, seasonal effects had virtually disappeared, and the evaporation rate was about 2 mm/day (0.08 in./day) in both summer and winter. By the 21st day, the evaporation rate was about 0.75 mm/day (0.03 in./day) in both summer and winter. Data were presented that illustrated a diurnal fluctuation in evaporation still persisting on the 37th day after irrigation.

DIRECT EVALUATION OF SOIL WATER FLUX AFTER IRRIGATION

Time-depth patterns of soil water flux in the 0- to 9-cm (0- to 3.5-in.) zone of a bare field soil were presented for four 24-hr periods at 3, 7, 16, and 37 days after irrigation by Jackson et al.[15] Their objective was to obtain a more direct evaluation of soil water flux than had previously been obtained.

The paper presented results from an experiment in which samples were taken a total of 48 times at 0.5-hr intervals for each of sixteen 24-hr periods. At 0.5 intervals, direct measurement was made of the evaporative flux at the surface using weighing lysimeters. From the 0-depth flux and water content data, flux values at eight soil depths at each measurement time were calculated directly.

Two weighing lysimeters and the surrounding 91 × 73 m (299 × 240 ft) plot were irrigated with about 10 cm (3.9 in.) of water. Soil water contents were measured in a 20 × 20 m (65.6 ft) subplot near the lysimeters. The error in smoothed lysimeter-determined evaporation rates was conservatively estimated to be ±0.02 mm/hr (±0.0008 in./hr); the estimate of overall error in soil water flux calculation was ±0.04 mm/hr (±0.0016 in./hr).

On the third day after irrigation, the soil water flux at 0 cm (evaporation) dominated the flux patterns for all depths. As the soil dried, the 0-cm flux decreased and the fluxes at the greater depths gradually became dominant on all of the first 16 days of measurement. Downward flux was observed below 1 to 3 cm (0.4 to 1.2 in.) during several hours between sunrise and early afternoon. One period of downward flux was observed for day 3; for subsequent days, 2 to 4 periods of downward flux were observed.

The data demonstrated the dynamic nature of soil water flux in the surface zone of a field soil subjected to diurnally-varying environmental conditions.

USE OF LYSIMETER TO SCHEDULE IRRIGATION AUTOMATICALLY

Phene and others[16,17] had shown that when a water reservoir is attached to a lysimeter system and the irrigation reservoir is weighed with the lysimeter and refilled automatically each day at a given time, automated measurements of crop evapotranspiration can be made without interruption, and the lysimeter can be used in a feedback mode to schedule irrigation automatically. Phene et al.[18] reported that the purposes were "(1) to develop design modifications which permit the simultaneous utilization of lysimeters as an evapotranspiration measuring device and as a closed-loop feedback automated irrigation controller and (2) to present the results of an experiment which substantiate the validity of the method for irrigation scheduling and evapotranspiration research, including generating a crop coefficient function for the irrigation season and maintaining steady state soil matric potential in the crop root zone."

Two lysimeters of the weighing type were used in this latter study. One was located in the middle of a 2-ha (5-acre) cultivated field (it is referred to as the crop lysimeter); the other was planted in grass and located in the middle of a 2-ha (5-acre) irrigated grass field (it is referred to as the grass or reference lysimeter) next to a weather station. The surface area of each of the lysimeters was 4 m^2 (43 ft^2) and the soil profile depth was 2 m (6.6 ft) overlying 0.3

Figure 2. Lysimeter planted in furrow-irrigated tomatoes. (Reprinted from Department of Land, Air, and Water Resources, University of California, Davis.)

m (0.98 ft) of fine sand. Each lysimeter rested on a sensitive electronic scale which was capable of measuring the total mass of approximately 20 Mg (44,000 lb) to the nearest 100 g (0.22 lb).

A water supply tank was attached to each of the lysimeters so that the mass of the daily irrigation volume was included in the mass of the lysimeter. Every hour on the hour, the data acquisition system recorded the weight of each lysimeter, calculated the mass change from the previous hour, and calculated the accumulated mass loss. When a mass change, a 4-kg (8.8-lb) mass loss, equivalent to 1.0 mm (0.04 in.), of cumulative evapotranspiration had occurred during the previous hour, the lysimeter was automatically irrigated with volume of water equal to the 1.0 mm (0.04 in.). Hourly evapotranspiration was calculated from the hourly mass changes without interruption, and the gradual increase in lysimeter mass with time represented an increase in crop fresh mass, while mass decrease during the day represented evapotranspiration. The rate of biomass accumulation was about 25 g/hr (0.055 lb/hr).

The lysimeter control system was used to control three different irrigation systems in 12 large field plots (three treatments replicated four times). Each plot [90 m (295 ft) × 17 m (56 ft)] was planted in tomatoes in 1984 and 1985, planted in cantaloupes in 1986, and planted in tomatoes in 1987. Daily evapotranspiration from the crop lysimeter, E_{tc}, and from the grass lysimeter, E_{to}, were used to generate a crop coefficient, K_c, expressed as: $K_c = E_{tc}/E_{to}$.

Results from four years of testing indicated that the automated weighing lysimeter as a feedback irrigation controller can be used to determine hourly crop evapotranspiration, establish a crop coefficient function, and schedule irrigations in surrounding fields irrigated under the same conditions as the lysimeter (subsurface drip irrigation) or under slightly different conditions (high frequency and low frequency surface drip irrigation). A lysimeter planted in furrow-irrigated tomatoes is shown in Figure 2.

REFERENCES

1. Kohnke, H., and F. R. Dreibelbis, "Lysimeters: A Survey and Discussion," in "A Survey and Discussion of Lysimeters and a Bibliography on Their Construction and Performance," USDA Misc. Pub. No. 372 (1940).
2. De la Hire, P. "Sur l'Origine des Rivieres," *Hist. de l'Acad. Roy. des Sci.* 1—6 (1703).
3. Sturtevant, E. L. "Waushakum Farm Experiments," *Sci. Farmer* 1:90 (1876).
4. Tanner, C. B. "Measurement of Evapotranspiration," in *Irrigation of Agricultural Lands,* R. M. Hagan, H. R. Haise, and T. W. Edminister, Eds., Am. Soc. of Agronomy Monograph No. 11, 1967, pp. 534—574.
5. Black, T. A., G. W. Thurtell, and C. B. Tanner. "Hydraulic Load-Cell Lysimeter, Construction, Calibration, and Tests," *Soil. Sci. Soc. Am. Proc.* 32:623—629 (1968).
6. Marek, T. H., A. D. Schneider, T. A. Howell, and L. L. Ebeling. "Design and Construction of Large Weighing Monolithic Lysimeters," *Trans. ASAE* 31:477—484 (1988).
7. Hatfield, J. L. "Methods of Estimating Evapotranspiration," *Irrigation of Agricultural Crops—Agronomy Monograph* (Madison, WI: ASA-CSSA-SSSA, 1990).
8. Harrold, L. L., and F. R. Dreibelbis. "Water Use by Crops as Determined by Weighing Monolith Lysimeters," *Soil Sci. Soc. Am. Proc.* 17:70—74 (1953).
9. Pruitt, W. O., and D. E. Angus. "Large Weighing Lysimeter for Measuring Evapotranspiration," *Trans. ASAE* 3:13—18 (1960).
10. van Bavel, C. H. M., and L. E. Myers. "An Automatic Weighing Lysimeter," *Agric. Eng.* 43:580—583, 587—588 (1962).
11. Dugas, W. A., D. R. Upchurch, and J. T. Ritchie. "A Weighing Lysimeter for Evapotranspiration and Root Measurements," *Agron. J.* 77:821—825 (1985).

12. Howell, T. A., R. L. McCormick, and C. J. Phene. "Design and Installation of Large Weighing Lysimeters," *Trans. ASAE* 28:106—112, 117 (1985).
13. Kruse, E. G., and C. M. U. Neale. "Sources of Error in Hydraulic Weighing Lysimeter Measurements," *Trans. ASAE* 32:81—96 (1989).
14. Kimball, B. A., and R. D. Jackson. "Seasonal Effects on Soil Drying After Irrigation," *Hydrol. Water Res. Ariz. Southwest* 1:85—98 (1971).
15. Jackson, R. D., B. A. Kimball, R. J. Reginato, and F. S. Nakayama. "Diurnal Soil-Water Evaporation: Time-Depth-Flux Patterns," *Soil Sci. Soc. Am. Proc.* 37:505—509 (1973).
16. Phene, C. J., et al. "Evapotranspiration and Crop Trickle Coefficient of Trickle Irrigated Tomatoes," *Third International Drip/Trickle Irrigation Congress*, ASAE Publisher 10—85:823—831 (1985).
17. Phene, C. J., et al. "Automatic Feedback Irrigation Scheduling and Control with a Weighing Lysimeter," *Proc. Agrimation* 2:135—147 (1986).
18. Phene, C. J., R. L. McCormick, K. R. Davis, J. D. Pierro, and D. W. Meek. "A Lysimeter Feedback Irrigation Controller System for Evapotranspiration Measurements and Real Time Irrigation Scheduling," *Trans. ASAE* 32:477—484 (1989).

CHAPTER 12

Evaporation Reduction by Various Means

INTRODUCTION

In Chapter 8, the use of monomolecular films in attempting to reduce evaporation was reviewed. Evaporation reduction by this means had averaged only about 20% in field studies.[1-3]

A more promising approach to evaporation reduction is the development and application of methods which reduce the energy available for evaporation.[3] These methods decrease evaporation either by reducing the solar energy entering the water or by reducing the transport of water vapor above the surface. A shade suspended above the surface of the water reduces the solar energy entering the water. The transport of water vapor has been reduced by wind baffles placed around, on, or above the water surface.

EFFECT OF REFLECTION OF SOLAR ENERGY ON EVAPORATION OF WATER

Beard and Gainer[4] studied the effect of solar energy reflectance on water evaporation. The results showed how ideal chemical films with different reflectance properties influence the amount of evaporation.

A search was made to find a monolayer or film-forming material (already colored or could be colored in some way) with good diffusion barrier capabilities. Tests were run to determine the percentage increase in solar reflection of these films and monolayers over that of a plain water surface. An analytical method was used to predict the changes in evaporation that would accompany various arbitrary changes in the reflectance of the water surface.

For maximum evaporation suppression, a monolayer or film should not only act as a diffusion barrier, but must also act to limit the amount of solar energy absorbed by the water surface. Only one monolayer-forming substance was found to significantly increase the solar energy reflectance — a special Union Carbide experimental silicone, S-1362-91-2. This film material, in addition to acting as a diffusion barrier and increasing the solar energy reflectance, also had good spreading properties and was extremely difficult to remove from a water surface.

REDUCTION OF EVAPORATION FROM WATER SURFACES BY THE USE OF REFLECTIVE SURFACES

Cooley and Myers[5] studied the efficiencies and economics of materials that they considered met the theoretical requirements of high reflectivity and emittance needed for efficient evaporation reduction from water surfaces. Of the more than 20 materials tested, the results for 8 of the more promising or practical were presented.

Myers and Frasier[6] investigated a number of floating reflective materials; their data showed that the water in the treated tank to be from 1 to 4.5°C cooler than in the open tank. Albedo (or reflectivity) over the treated tanks was from 10 to 48% greater than over the open tank, and the evaporation was reduced from 17 to 45%, depending on the treatment. They concluded that the floating reflective materials could reduce annual evaporation losses by 40%.

Experiments were conducted with two sets of metal stock tanks selected to represent a wide range of field conditions. The first set was four insulated buried tanks; to simulate small ponds or reservoirs, each had an inner tank measuring 2.1 m (6.9 ft) in diameter and 0.6 m (2 ft) in depth, and an outer tank measuring 2.7 m (8.9 ft) in diameter and 0.9 m (3 ft) in depth. The rims of the tanks were about 8 cm (3 in.) above the ground, and the surrounding area for at least 20 m (66 ft) was covered with irrigated lawn or bare soil. The spaces between the inner and outer tanks were filled with perlite ore as an insulator. The second set was six uninsulated tanks; to simulate steel and concrete stock tanks, each measured 2.7 m (8.9 ft) in diameter and 0.9 m (3 ft) in depth. These tanks and the other rows of three tanks were spaced about 50 cm (20 in.) apart. The surrounding area was bare soil, except on the south side where a 14-m (46-ft) × 30-m (98-ft) pond was within 4 m (13 ft) to 5 m (16 ft) of the tanks.

One tank in each set was left untreated for comparison with the treated tanks. Before treatment, the tanks were calibrated; the four insulated tanks had variations of less than 3%, the six exposed-wall tanks had variations of up to 8%, and the middle two tanks evaporated 8% less than the end tanks (indicating a shading effect). These effects were taken into account when calculating evaporation reduction efficiencies of the treatments.

Evaporation measurements were determined from point gauge readings taken about twice weekly. Water temperatures were recorded occasionally with a standard mercury thermometer near the surface and beneath the water at approximately $^1/_2$ the depth of the tank.

The eight reflective materials included in the study were: asphalt-concrete blocks, white butyl rubber, SSP foamed butyl rubber, Mini-Vaps® (a commercial white plastic material in the shape of a plus sign), Styrofoam,® foamed wax blocks, white foamed wax with melting point in the 120 to 125°C range, and white foamed wax with melting point in the 120 to 135°C range. In all cases, the lighter the cover color the greater the efficiency in evaporation reduction. The Styrofoam,® white butyl, SSP foamed butyl, and the wax blocks were the most efficient materials, in that order.

Figure 1. Foamed wax blocks on exposed-wall evaporation tank. (Reprinted from U.S. Water Conservation Laboratory, U.S. Department of Agriculture, Phoenix, Arizona.)

The eight materials tested reduced the annual evaporation losses 23 to 87%. The best results were obtained from the foamed wax blocks and a 120 to 125°C melting point continuous wax layer on the buried and exposed-wall tanks, respectively. The authors suggested that the physical characteristics of the storage facility should also be considered in determining which cover material to use. Foamed wax blocks on an exposed-wall tank are shown in Figure 1.

EVAPORATION REDUCTION BY REDUCTION OF AVAILABLE SOLAR ENERGY

Cooley[3] observed that water lost by evaporation from open surfaces may equal or exceed that used beneficially, and that conserving water contained in existing facilities is, in some situations, the most economical means of providing adequate water supplies. Various energy-reducing methods for evaporation reduction can be classified as: (1) changing the color of the water; (2) using wind barriers; (3) shading the water surface; and (4) using floating covers. Table 1 summarizes the results in the literature for these methods. Of these methods, floating covers have been most widely used.

Table 1 Evaporation Reduction Achieved by Various Energy-Reducing Methods

Method		Area of water surface covered (%)	Evaporation reduction (%)	Reference
1.	Changing the water color:			
	Dye in water	100	6—9	7
	Shallow, colored pans	100	35—50[a]	8
2.	Using wind barriers:			
	Baffles	—	11	9
3.	Shading the water surface			
	Plastic mesh	47	44	9
	Blue polylaminated plastic sheeting	100	90	10
4.	Floating covers:			
	Perlite ore	78	19	11
	Polystyrene beads	78	39	6
	Wax blocks	78	64	5
	White spheres	78	78	9
	White butyl sheets	86	77	12
	Polystyrene sheets	80	79	12
	Polystyrene rafts	100	95	13
	Continuous wax	100	87	5
	Foamed rubber	95	90	14

Source: Cooley.[3]

[a] Evaporation from white pan compared with that from black pan.

REDUCTION OF EVAPORATION BY THE USE OF FLOATING COVERS

Recognizing that floating covers needed full evaluation, Cooley[3] made a long-term study to determine the effects of long-term exposure on the evaporation reduction efficiency of several of the most promising floating materials and to determine the cost of water saved when material and maintenance costs for long-term periods are considered. The study was made for conditions similar to actual field use on farms, range tanks, or ponds.

A large increase in reflectance created by floating reflective materials on the water can significantly reduce the evaporation rate. It is therefore desirable to make the surface of the reflecting material as light-colored as possible. Some floating covers also act as a physical barrier to evaporation.

All of the floating covers were studied on a set of six metal stock tanks (2.7 m (8.9 ft) in diameter and 0.9 m (3 ft) in depth) located about 50 km (31 mi) northeast of Phoenix, Arizona. The tanks were placed on top of the ground with their sides exposed. One tank was left untreated for comparison, and all of the tanks were calibrated before treatment, showing variations of up to 8% in the six tanks. Evaporation measurements were determined from point gauge readings taken about twice a week.

The five materials included in the study were: (1) white foamed wax blocks of high melting point (150 to 165°C), covering about 60% of the surface area;

(2) continuous foamed wax (melting point 120 to 135°C) which melted about 6 months after installation and formed a gray-brown continuous wax layer about 0.6 cm (0.24 in.) thick, covering 100% of the surface area; (3) continuous foamed wax (melting point 120 to 125°C) which melted within a month after installation and formed a gray-brown continuous wax layer about 0.6 cm (0.24 in.) thick, covering 100% of the surface layer; (4) continuous wax blocks (melting point 120 to 125°C), installed with a layer of carbon black on top, which melted within six weeks and formed a gray-black continuous layer about 0.4 cm (0.16 in.) thick, covering 100% of the surface area; and (5) foamed rubber or sponge (originally black, but turned to gray-brown as it aged and became covered with dust) made of low-density closed-cell ethylene propylene diene monomer (EPDM) synthetic rubber about 0.6 cm (0.24 in.) thick.

Average evaporation reduction efficiency ranged from 36% over an 8-year period for the foamed wax blocks which covered about 60% of the water surface to 84% over a 4-year period for the foamed rubber cover which covered about 95% of the water surface. Although more water was saved by the foamed rubber cover in 4 years than by the foamed wax blocks in 8 years, the continuous wax covers of 120 to 125°C melting point provided the lowest cost water. All of the floating covers tested saved water for less than 10% of the cost of hauling water to range stock tanks.

Foamed water covers have been used successfully in hot and cold climates and on larger tanks (300,000-L, 79,250-gal) with vertical-walled steel tanks about 11 m (36 ft) in diameter.

EVAPORATION REDUCTION IN STOCK TANKS

Khan and Issac[15] evaluated the performance and economics of different locally available materials in reducing stock tank evaporation in arid regions in India.

Evaporation reduction studies were conducted in seven cemented tanks partially sunk in the ground, located at a research farm at Jodhpur, India. The tanks were 200 cm × 200 cm × 100 cm deep (79 in. × 79 in. × 39 in. deep) and spaced 200 cm (79 in.) apart. The walls of the tanks were 30 cm (12 in.) thick and extended 20 cm (7.9 in.) above the ground level; the bottoms were 20 cm (7.9 in.) thick. The tanks were surrounded by bare soil. One tank was left untreated for comparison. The tanks were filled to a level of 95 cm (37 in.), and were calibrated for a period of 30 days before imposing the treatments. Variation in the water loss was 2.5% for the calibration.

A point gauge installed in each of the tanks was used to make evaporation measurements at 9:30 a.m. and 4:30 p.m. A standard rain gauge near the tanks was used for recording the rainfall. A digital temperature meter was used to record water temperature at a depth of 1 cm (0.4 in.), at $^1/_2$ the depth of the tank, and at the bottom level depth in the treated and untreated tanks.

All of the materials evaluated were white or of a light color. Covers acted as reflectors and as vapor barriers. The materials evaluated for shading the water surface and for floating covers were:

Shading the Water Surface

- A 0.25-mm (0.01-in.) thick polyethene sheet shading the complete water surface. The edges of the sheet were anchored and buried in the ground around the tank.

Floating Covers

- A 0.25-mm (0.01-in.) thick white polyethene sheet mounted on a bamboo frame, covering 75% of the water surface area.
- A 20-cm (0.79-in.) thick foam rubber sheet painted white on top, covering 90% of the water area.
- 25-mm (0.98-in.) thick white polystyrene sheets (100 × 50 cm, 39.3 in. × 20 in.) were placed end to end with overlapping joints, covering 98% of the area.
- Bamboo averaging about 25 mm (0.98 in.) in diameter and 190 cm (74.8 in.) long, tied together to form a continuous layer, covering 88% of the surface area.
- The stem of *Sacchrum munja,* about 6 mm (0.2 in.) in diameter, flowed together to form a continuous layer, covering 90% of the surface area.

The floating polyethene sheet covering 75% of the water surface reduced evaporation by 66%. The foam rubber sheet, painted white to increase reflectance, covered 90% of the area and reduced evaporation by an average of 74%. The polystyrene sheet formed a complete vapor barrier over 98% of the water surface area and reduced evaporation by 82% over a 19-month period. The bamboo covered 98% of the water surface, and evaporation was reduced by an average of 54%. During the study period, the bamboo stems gradually absorbed water and gained weight, allowing water through the space between stems and causing a reduction of efficiency of 19% at the end of the study period. The stems of *S. munja* formed a vapor barrier over 90% of the water surface and reduced evaporation by over 50% for the first two months. It suffered degradation by solar radiation, absorbed water, and was coated with algae and dust, reducing its efficiency to 26% at the end of the study.

Monthly average water temperatures at half-tank level in treated and untreated tanks were compared; water in the treated tanks was 1 to 5°C cooler than in the open tank. Water temperature at the bottom level in treated tanks remained constant within 0.5°C throughout the day, indicating that convective heat transfer through the tank bottoms was negligible.

Shading the water surface with polyethene was the least expensive method. The floating polystyrene sheets and polyethene sheets were considerably more economical than any of the other floating materials tested.

REDUCTION OF EVAPORATION BY FLOATING RUBBER OR SPONGE TANK COVERS

In areas in which water supplies are limited and evaporation rates are high, floating foam rubber tank covers can cut evaporation by 80 to 90%.[16]

The floating foam rubber or sponge covers were made of low density, closed-cell (EPDM) synthetic rubber, weighing about 1 lb for 6 ft^2 of $^1/_4$ in.-thick sheeting. A 4 ft width was the most practical size. A foam rubber cover on a water storage tank is shown in Figure 2.

EFFECTS OF LILY PADS ON EVAPORATION

Cooley and Idso[17] studied the effects of lily pads on evaporation. For a period of about a month, they measured daily evaporation from a pair of sunken evaporation tanks that were identical in all respects, except that one of them had four common lily plants growing within it. The lily pads covered about 18% of the surface.

The tanks consisted of an outer framework measuring 2.7 m (8.9 ft) in diameter and 0.9 m (3.0 ft) in depth, and an inner container measuring 2.1 m (6.9 ft) in diameter and 0.6 m (2.0 ft) in depth. The smaller container was placed within the larger container with their top rims protruding about 2 to 3 cm (0.8 to 1.2 in.) above the level of the ground. The space between the two containers was filled with perlite ore.

Evaporation determinations were made every 30 min from water level measurements made by proximity sensing probes. Meteorological measurements, including the net radiation flux to each tank, were made at the same 30-min interval. The evaporative water loss from the tank with the lily pads floating on the water surface was 2.9% less than that from the open-water tank. This reduction was matched by a 2.7% reduction in measured radiation over the tank with lily pads. Evaporation from the surface area covered by lily pads was calculated to be about 84% of that occurring from open water.

Figure 2. A foam rubber cover on a water storage tank. (Reprinted from U.S. Water Conservation Laboratory, U.S. Department of Agriculture, Phoenix, Arizona.)

REFERENCES

1. Cluff, C. B. "Evaporation Reduction Investigations Relating to Small Reservoirs," *Technical Bulletin No. 177,* Arizona Agricultural Experiment Station, Oct., 1966, pp. 1—47.
2. Frasier, G. W., and L. W. Myers. "Stable Alkanol Dispersion to Reduce Evaporation," *J. Irrig. Drainage Div. ASCE* 94:79—89 (1968).
3. Cooley, K. R. "Evaporation Reduction: Summary of Long-Term Tank Studies," *J. Irrig. Drainage Div. ASCE* 109:89—98 (1983).
4. Beard, J. T., and J. L. Gainer. "Influence of Solar Radiation Reflectance on Water Evaporation," *J. Geophys. Res.* 75:5155—5163 (1970).
5. Cooley, K. R., and L. E. Myers. "Evaporation Reduction with Reflective Covers," *J. Irrig. Drainage Div. ASCE* 99(IR3):353—363 (1973).
6. Myers, L. E., and G. W. Frasier. "Evaporation Reduction with Floating Granular Materials," *J. Irrig. Drainage Div. ASCE* 96(IR4):425—436 (1970).
7. Keyes, C. G., and N. N. Gunaji. "Effect of Dye on Solar Evaporation of Brine," *International Union of Geodesy and Geophysics General Assembly of Bern Publication No. 78,* International Association of Scientific Hydrology, 1967, pp. 338—347.
8. Yu, S. L., and W. Brutsaert. "Evaporation From Very Shallow Pans," *J. Appl. Meteor.* 6:265—271 (1967).
9. Crow, F. R., and H. L. Manges. "Comparison of Chemical and Non-Chemical Techniques for Suppressing Evaporation from Small Reservoirs," *Trans. ASAE* 10:172—174 (1967).
10. Drew, W. M. "Evaporation Control—A Comparative Study of Six Evaporation Restriction Media," *Aqua,* Jan., 1972, pp. 23—26.
11. Cooley, K. R., and C. B. Cluff. "Reducing Pond Evaporation with Perlite Ore," *J. Irrig. Drainage Div. ASCE* 98(IR2):255—266 (1972).
12. Cooley, K. R. "Energy Relationships in the Design of Floating Covers for Evaporation Control," *Water Resour. Res.* 6:717—727 (1970).
13. Cluff, C. B. "Patchwork Quilt Halts Water Evaporation Loss," *California Farmer* 237:(1972), 18 pp.
14. Dedrick, A. R., T. D. Hansen, and W. R. Williamson. "Floating Sheets of Foam Rubber for Reducing Stock Tank Evaporation," *J. Range Manage.* 26:404—406 (1973).
15. Khan, M. A., and V. C. Issac. "Evaporation Reduction in Stock Tanks for Increasing Water Supplies," *J. Hydrol.* 119:21—29 (1990).
16. Dedrick, A. R. "Foam Rubber Covers for Controlling Water Storage Tank Evaporation," University of Arizona College of Agriculture Cooperative Extension Service Misc. Rep. No. 2 (1976).
17. Cooley, K. R., and S. B. Idso. "Effects of Lily Pads on Evaporation," *Water Resour. Res.* 16:605—606 (1980).

Index